Selected Titles in This Series

92 **Fan R. K. Chung,** Spectral graph theory, 1997

91 **J. P. May et al.,** Equivariant homotopy and cohomology theory, ded... of Robert J. Piacenza, 1996

90 **John Roe,** Index theory, coarse geometry, and topology of manifolds, 1996

89 **Clifford Henry Taubes,** Metrics, connections and gluing theorems, 1996

88 **Craig Huneke,** Tight closure and its applications, 1996

87 **John Erik Fornæss,** Dynamics in several complex variables, 1996

86 **Sorin Popa,** Classification of subfactors and their endomorphisms, 1995

85 **Michio Jimbo and Tetsuji Miwa,** Algebraic analysis of solvable lattice models, 1994

84 **Hugh L. Montgomery,** Ten lectures on the interface between analytic number theory and harmonic analysis, 1994

83 **Carlos E. Kenig,** Harmonic analysis techniques for second order elliptic boundary value problems, 1994

82 **Susan Montgomery,** Hopf algebras and their actions on rings, 1993

81 **Steven G. Krantz,** Geometric analysis and function spaces, 1993

80 **Vaughan F. R. Jones,** Subfactors and knots, 1991

79 **Michael Frazier, Björn Jawerth, and Guido Weiss,** Littlewood-Paley theory and the study of function spaces, 1991

78 **Edward Formanek,** The polynomial identities and variants of $n \times n$ matrices, 1991

77 **Michael Christ,** Lectures on singular integral operators, 1990

76 **Klaus Schmidt,** Algebraic ideas in ergodic theory, 1990

75 **F. Thomas Farrell and L. Edwin Jones,** Classical aspherical manifolds, 1990

74 **Lawrence C. Evans,** Weak convergence methods for nonlinear partial differential equations, 1990

73 **Walter A. Strauss,** Nonlinear wave equations, 1989

72 **Peter Orlik,** Introduction to arrangements, 1989

71 **Harry Dym,** J contractive matrix functions, reproducing kernel Hilbert spaces and interpolation, 1989

70 **Richard F. Gundy,** Some topics in probability and analysis, 1989

69 **Frank D. Grosshans, Gian-Carlo Rota, and Joel A. Stein,** Invariant theory and superalgebras, 1987

68 **J. William Helton, Joseph A. Ball, Charles R. Johnson, and John N. Palmer,** Operator theory, analytic functions, matrices, and electrical engineering, 1987

67 **Harald Upmeier,** Jordan algebras in analysis, operator theory, and quantum mechanics, 1987

66 **G. Andrews,** q-Series: Their development and application in analysis, number theory, combinatorics, physics and computer algebra, 1986

65 **Paul H. Rabinowitz,** Minimax methods in critical point theory with applications to differential equations, 1986

64 **Donald S. Passman,** Group rings, crossed products and Galois theory, 1986

63 **Walter Rudin,** New constructions of functions holomorphic in the unit ball of C^n, 1986

62 **Béla Bollobás,** Extremal graph theory with emphasis on probabilistic methods, 1986

61 **Mogens Flensted-Jensen,** Analysis on non-Riemannian symmetric spaces, 1986

60 **Gilles Pisier,** Factorization of linear operators and geometry of Banach spaces, 1986

59 **Roger Howe and Allen Moy,** Harish-Chandra homomorphisms for p-adic groups, 1985

58 **H. Blaine Lawson, Jr.,** The theory of gauge fields in four dimensions, 1985

57 **Jerry L. Kazdan,** Prescribing the curvature of a Riemannian manifold, 1985

(*Continued in the back of this publication*)

Spectral Graph Theory

Conference Board of the Mathematical Sciences

CBMS

Regional Conference Series in Mathematics

Number 92

Spectral Graph Theory

Fan R. K. Chung

Published for the
Conference Board of the Mathematical Sciences
by the
American Mathematical Society
Providence, Rhode Island
with support from the
National Science Foundation

CBMS Conference on Recent Advances
in Spectral Graph Theory
held at California State University at Fresno,
June 6–10, 1994

Research partially supported by the
National Science Foundation

1991 *Mathematics Subject Classification.* Primary 05-XX.

Library of Congress Cataloging-in-Publication Data
Chung, Fan R. K.
    Spectral graph theory / Fan R. K. Chung.
      p. cm.—(Regional conference series in mathematics, ISSN 0160-7642 ; no. 92)
    "CBMS Conference on Recent Advances in Spectral Graph Theory held at California State
University at Fresno, June 6–10, 1994"—T.p. verso.
    Includes bibliographical references.
    ISBN 0-8218-0315-8 (soft : alk. paper)
    1. Graph theory—Congresses. 2. Eigenvalues—Congresses. I. CBMS Conference on Recent
Advances in Spectral Graph Theory (1994 : California State University at Fresno) II. Title.
III. Series.
QA1.R33 no. 92
[QA166]
511′.5—dc21   96-45112
                                                                                           CIP

Contents

Preface   xi

Chapter 1.   Eigenvalues and the Laplacian of a graph   1

   1.1.   Introduction   1

   1.2.   The Laplacian and eigenvalues   2

   1.3.   Basic facts about the spectrum of a graph   6

   1.4.   Eigenvalues of weighted graphs   11

   1.5.   Eigenvalues and random walks   14

Chapter 2.   Isoperimetric problems   23

   2.1.   History   23

   2.2.   The Cheeger constant of a graph   24

   2.3.   The edge expansion of a graph   25

   2.4.   The vertex expansion of a graph   29

   2.5.   A characterization of the Cheeger constant   32

   2.6.   Isoperimetric inequalities for cartesian products   36

Chapter 3.   Diameters and eigenvalues   43

   3.1.   The diameter of a graph   43

   3.2.   Eigenvalues and distances between two subsets   45

   3.3.   Eigenvalues and distances among many subsets   49

   3.4.   Eigenvalue upper bounds for manifolds   50

Chapter 4.   Paths, flows, and routing   59

4.1.   Paths and sets of paths   59

4.2.   Flows and Cheeger constants   60

4.3.   Eigenvalues and routes with small congestion   62

4.4.   Routing in graphs   64

4.5.   Comparison theorems   68

Chapter 5.   Eigenvalues and quasi-randomness   73

5.1.   Quasi-randomness   73

5.2.   The discrepancy property   75

5.3.   The deviation of a graph   81

5.4.   Quasi-random graphs   85

Chapter 6.   Expanders and explicit constructions   91

6.1.   Probabilistic methods versus explicit constructions   91

6.2.   The expanders   92

6.3.   Examples of explicit constructions   97

6.4.   Applications of expanders in communication networks   102

6.5.   Constructions of graphs with small diameter and girth   105

6.6.   Weighted Laplacians and the Lovász ϑ function 107

Chapter 7.   Eigenvalues of symmetrical graphs   113

7.1.   Symmetrical graphs   113

7.2.   Cheeger constants of symmetrical graphs   114

7.3.   Eigenvalues of symmetrical graphs   116

7.4.   Distance transitive graphs   118

7.5.   Eigenvalues and group representation theory   121

7.6.   The vibrational spectrum of a graph   123

Chapter 8.   Eigenvalues of subgraphs with boundary conditions 127

8.1.   Neumann eigenvalues and Dirichlet eigenvalues   127

8.2.   The Neumann eigenvalues of a subgraph   128

8.3.   Neumann eigenvalues and random walks   130

8.4.   Dirichlet eigenvalues   132

8.5.   A matrix-tree theorem and Dirichlet eigenvalues   133

8.6.   Determinants and invariant field theory   135

Chapter 9.   Harnack inequalities   139

9.1.   Eigenfunctions   139

9.2.   Convex subgraphs of homogeneous graphs   140

9.3.   A Harnack inequality for homogeneous graphs   142

9.4.   Harnack inequalities for Dirichlet eigenvalues   144

9.5.   Harnack inequalities for Neumann eigenvalues   146

9.6.   Eigenvalues and diameters   148

Chapter 10.   Heat kernels   149

10.1.   The heat kernel of a graph and its induced subgraphs   149

10.2.   Basic facts on heat kernels   150

10.3.   An eigenvalue inequality   152

10.4.   Heat kernel lower bounds   154

10.5.   Matrices with given row and column sums   160

10.6.   Random walks and the heat kernel   165

Chapter 11.   Sobolev inequalities   167

11.1.   The isoperimetric dimension of a graph   167

11.2.   An isoperimetric inequality   169

11.3.   Sobolev inequalities   172

11.4.   Eigenvalue bounds   174

11.5.   Generalizations to weighted graphs and subgraphs   179

Chapter 12.   Advanced techniques for random walks on graphs   181

12.1.   Several approaches for bounding convergence   181

12.2.   Logarithmic Sobolev inequalities   184

12.3.   A comparison theorem for the log-Sobolev constant   189

12.4.   Logarithmic Harnack inequalities   191

12.5.   The isoperimetric dimension and the Sobolev inequality   196

Bibliography   201

Index   210

Preface

This monograph is an intertwined tale of eigenvalues and their use in unlocking a thousand secrets about graphs. The stories will be told — how the spectrum reveals fundamental properties of a graph, how spectral graph theory links the discrete universe to the continuous one through geometric, analytic and algebraic techniques, and how, through eigenvalues, theory and applications in communications and computer science come together in symbiotic harmony. Since spectral graph theory has been evolving very rapidly, the above goals can only be partially fulfilled here. For example, more advanced material on the heat kernel will be treated in a later publication.

This book is based on ten lectures given at the CBMS workshop on spectral graph theory in June 1994 at Fresno State University. Additional chapters were included on subgraphs with boundary conditions and on advanced techniques for random walks. I wish to thank S.T. Yau for introducing me to powerful ideas from spectral geometry. The last few chapters are mainly based on our collaborative work on geometry of graphs. Special thanks are due to Ron Graham, without whose encouragement this book would never have existed. In the course of writing, I have greatly benefitted from numerous suggestions and valuable comments of Noga Alon, Andy Woldar, Felix Lazebnik, David Gillman, Elizabeth Wilmer, Prasad Tetali and Herb Wilf. I would also like to acknowledge the support of National Science Foundation through Grant No. DMS 95-04834.

Like all authors, I would hope that these lecture notes are completely error-free. However, to be realistic, I plan to maintain an errata list on my home page *http://www.math.upenn.edu/~chung*. Naturally, I will be grateful for any contribution to this list.

Fan Chung

September, 1996

CHAPTER 1

Eigenvalues and the Laplacian of a graph

1.1. Introduction

Spectral graph theory has a long history. In the early days, matrix theory and linear algebra were used to analyze adjacency matrices of graphs. Algebraic methods are especially effective in treating graphs which are regular and symmetric. Sometimes, certain eigenvalues have been referred to as the "algebraic connectivity" of a graph [126]. There is a large literature on algebraic aspects of spectral graph theory, well documented in several surveys and books, such as Biggs [25], Cvetković, Doob and Sachs [90, 91], and Seidel [225].

In the past ten years, many developments in spectral graph theory have often had a geometric flavor. For example, the explicit constructions of expander graphs, due to Lubotzky-Phillips-Sarnak [194] and Margulis [196], are based on eigenvalues and isoperimetric properties of graphs. The discrete analogue of the Cheeger inequality has been heavily utilized in the study of random walks and rapidly mixing Markov chains [225]. New spectral techniques have emerged and they are powerful and well-suited for dealing with general graphs. In a way, spectral graph theory has entered a new era.

Just as astronomers study stellar spectra to determine the make-up of distant stars, one of the main goals in graph theory is to deduce the principal properties and structure of a graph from its graph spectrum (or from a short list of easily computable invariants). The spectral approach for general graphs is a step in this direction. We will see that eigenvalues are closely related to almost all major invariants of a graph, linking one extremal property to another. There is no question that eigenvalues play a central role in our fundamental understanding of graphs.

The study of graph eigenvalues realizes increasingly rich connections with many other areas of mathematics. A particularly important development is the interaction between spectral graph theory and differential geometry. There is an interesting analogy between spectral Riemannian geometry and spectral graph theory. The concepts and methods of spectral geometry bring useful tools and crucial insights to the study of graph eigenvalues, which in turn lead to new directions and results in spectral geometry. Algebraic spectral methods are also very useful, especially for extremal examples and constructions. In this book, we take a broad approach with emphasis on the geometric aspects of graph eigenvalues, while including the algebraic aspects as well. The reader is not required to have special background in geometry, since this book is almost entirely graph-theoretic.

From the start, spectral graph theory has had applications to chemistry [27]. Eigenvalues were associated with the stability of molecules. Also, graph spectra arise naturally in various problems of theoretical physics and quantum mechanics, for example, in minimizing energies of Hamiltonian systems. The recent progress on expander graphs and eigenvalues was initiated by problems in communication networks. The development of rapidly mixing Markov chains has intertwined with advances in randomized approximation algorithms. Applications of graph eigenvalues occur in numerous areas and in different guises. However, the underlying mathematics of spectral graph theory through all its connections to the pure and applied, the continuous and discrete, can be viewed as a single unified subject. It is this aspect that we intend to cover in this book.

1.2. The Laplacian and eigenvalues

Before we start to define eigenvalues, some explanations are in order. The eigenvalues we consider throughout this book are not exactly the same as those in Biggs [25] or Cvetković, Doob and Sachs [90]. Basically, the eigenvalues are defined here in a general and "normalized" form. Although this might look a little complicated at first, our eigenvalues relate well to other graph invariants for general graphs in a way that other definitions (such as the eigenvalues of adjacency matrices) often fail to do. The advantages of this definition are perhaps due to the fact that it is consistent with the eigenvalues in spectral geometry and in stochastic processes. Many results which were only known for regular graphs can be generalized to all graphs. Consequently, this provides a coherent treatment for a general graph. For definitions and standard graph-theoretic terminology, the reader is referred to [31].

In a graph G, let d_v denote the degree of the vertex v. We first define the Laplacian for graphs without loops and multiple edges (the general weighted case with loops will be treated in Section 1.4). To begin, we consider the matrix L, defined as follows:

$$L(u,v) = \begin{cases} d_v & \text{if } u = v, \\ -1 & \text{if } u \text{ and } v \text{ are adjacent,} \\ 0 & \text{otherwise.} \end{cases}$$

Let T denote the diagonal matrix with the (v,v)-th entry having value d_v. The *Laplacian* of G is defined to be the matrix

$$\mathcal{L}(u,v) = \begin{cases} 1 & \text{if } u = v \text{ and } d_v \neq 0, \\ -\dfrac{1}{\sqrt{d_u d_v}} & \text{if } u \text{ and } v \text{ are adjacent,} \\ 0 & \text{otherwise.} \end{cases}$$

We can write

$$\mathcal{L} = T^{-1/2} L T^{-1/2}$$

with the convention $T^{-1}(v,v) = 0$ for $d_v = 0$. We say v is an isolated vertex if $d_v = 0$. A graph is said to be nontrivial if it contains at least one edge.

$\mathcal{L}$ can be viewed as an operator on the space of functions $g : V(G) \to \mathbb{R}$ which satisfies

$$\mathcal{L}g(u) = \frac{1}{\sqrt{d_u}} \sum_{\substack{v \\ u \sim v}} \left(\frac{g(u)}{\sqrt{d_u}} - \frac{g(v)}{\sqrt{d_v}} \right)$$

When G is k-regular, it is easy to see that

$$\mathcal{L} = I - \frac{1}{k} A,$$

where A is the adjacency matrix of G,(i. e., $A(x,y) = 1$ if x is adjacent to y, and 0 otherwise,) and I is an identity matrix. All matrices here are $n \times n$ where n is the number of vertices in G.

For a general graph, we have

$$\begin{aligned} \mathcal{L} &= T^{-1/2} L T^{-1/2} \\ &= I - T^{-1/2} A T^{-1/2}. \end{aligned}$$

We note that $\mathcal{L}$ can be written as

$$\mathcal{L} = S \, S^*,$$

where S is the matrix whose rows are indexed by the vertices and whose columns are indexed by the edges of G such that each column corresponding to an edge $e = \{u, v\}$ has an entry $1/\sqrt{d_u}$ in the row corresponding to u, an entry $-1/\sqrt{d_v}$ in the row corresponding to v, and has zero entries elsewhere. (As it turns out, the choice of signs can be arbitrary as long as one is positive and the other is negative.) Also, S^* denotes the transpose of S.

For readers who are familiar with terminology in homology theory, we remark that S can be viewed as a "boundary operator" mapping "1-chains" defined on edges (denoted by C_1) of a graph to "0-chains" defined on vertices (denoted by C_0). Then, S^* is the corresponding "coboundary operator" and we have

$$C_1 \underset{S^*}{\overset{S}{\underset{\longleftarrow}{\longrightarrow}}} C_0$$

Since $\mathcal{L}$ is symmetric, its eigenvalues are all real and non-negative. We can use the variational characterizations of those eigenvalues in terms of the Rayleigh quotient of $\mathcal{L}$ (see, e.g. [163]). Let g denote an arbitrary function which assigns to each vertex v of G a real value $g(v)$. We can view g as a column vector. Then

$$(1.1) \qquad \begin{aligned} \frac{\langle g, \mathcal{L}g \rangle}{\langle g, g \rangle} &= \frac{\langle g, T^{-1/2} L T^{-1/2} g \rangle}{\langle g, g \rangle} \\ &= \frac{\langle f, L f \rangle}{\langle T^{1/2} f, T^{1/2} f \rangle} \\ &= \frac{\sum_{u \sim v} (f(u) - f(v))^2}{\sum_v f(v)^2 d_v} \end{aligned}$$

where $g = T^{1/2}f$ and $\sum_{u \sim v}$ denotes the sum over all unordered pairs $\{u, v\}$ for which u and v are adjacent. Here $\langle f, g \rangle = \sum_x f(x)g(x)$ denotes the standard inner product in $\mathbb{R}^n$. The sum $\sum_{u \sim v}(f(u) - f(v))^2$ is sometimes called the *Dirichlet sum* of G and the ratio on the left-hand side of (1.1) is often called the *Rayleigh quotient*. (We note that we can also use the inner product $\langle f, g \rangle = \sum \overline{f(x)}g(x)$ for complex-valued functions.)

From equation (1.1), we see that all eigenvalues are non-negative. In fact, we can easily deduce from equation (1.1) that 0 is an eigenvalue of $\mathcal{L}$. We denote the eigenvalues of $\mathcal{L}$ by $0 = \lambda_0 \leq \lambda_1 \leq \cdots \leq \lambda_{n-1}$. The set of the λ_i's is usually called the *spectrum* of $\mathcal{L}$ (or the spectrum of the associated graph G.) Let $\mathbf{1}$ denote the constant function which assumes the value 1 on each vertex. Then $T^{1/2}\mathbf{1}$ is an eigenfunction of $\mathcal{L}$ with eigenvalue 0. Furthermore,

(1.2)
$$\lambda_G = \lambda_1 = \inf_{f \perp T\mathbf{1}} \frac{\displaystyle\sum_{u \sim v}(f(u) - f(v))^2}{\displaystyle\sum_v f(v)^2 d_v}.$$

The corresponding eigenfunction is $g = T^{1/2}f$ as in (1.1). It is sometimes convenient to consider the nontrivial function f achieving (1.2), in which case we call f a *harmonic eigenfunction* of $\mathcal{L}$.

The above formulation for λ_G corresponds in a natural way to the eigenvalues of the Laplace-Beltrami operator for Riemannian manifolds:

$$\lambda_M = \inf \frac{\displaystyle\int_M |\nabla f|^2}{\displaystyle\int_M |f|^2},$$

where f ranges over functions satisfying

$$\int_M f = 0.$$

We remark that the corresponding measure here for each edge is 1 although in the general case for weighted graphs the measure for an edge is associated with the edge weight (see Section 1.4.) The measure for each vertex is the degree of the vertex. A more general notion of vertex weights will be considered in Section 2.5.

We note that (1.2) has several different formulations:

$$(1.3) \qquad \lambda_1 = \inf_f \sup_t \frac{\displaystyle\sum_{u \sim v}(f(u) - f(v))^2}{\displaystyle\sum_v (f(v) - t)^2 d_v}$$

$$(1.4) \qquad = \inf_f \frac{\displaystyle\sum_{u \sim v}(f(u) - f(v))^2}{\displaystyle\sum_v (f(v) - \bar{f})^2 d_v},$$

where

$$\bar{f} = \frac{\displaystyle\sum_v f(v) d_v}{\text{vol } G},$$

and vol G denotes the volume of the graph G, given by

$$\text{vol } G = \sum_v d_v.$$

By substituting for $\bar{f}$ and using the fact that $N \displaystyle\sum_{i=1}^{N}(a_i - a)^2 = \sum_{i<j}(a_i - a_j)^2$ for $a = \displaystyle\sum_{i=1}^{N} a_i/N$, we have the following expression (which generalizes the one in [126]):

$$(1.5) \qquad \lambda_1 = \text{vol } G \inf_f \frac{\displaystyle\sum_{u \sim v}(f(u) - f(v))^2}{\displaystyle\sum_{u,v}(f(u) - f(v))^2 d_u d_v},$$

where $\displaystyle\sum_{u,v}$ denotes the sum over all unordered pairs of vertices u, v in G. We can characterize the other eigenvalues of $\mathcal{L}$ in terms of the Rayleigh quotient. The largest eigenvalue satisfies:

$$(1.6) \qquad \lambda_{n-1} = \sup_f \frac{\displaystyle\sum_{u \sim v}(f(u) - f(v))^2}{\displaystyle\sum_v f^2(v) d_v}.$$

For a general k, we have

$$(1.7) \qquad \lambda_k = \inf_f \sup_{g \in P_{k-1}} \frac{\displaystyle\sum_{u \sim v}(f(u) - f(v))^2}{\displaystyle\sum_v (f(v) - g(v))^2 d_v}$$

$$(1.8) \qquad = \inf_{f \perp T P_{k-1}} \frac{\displaystyle\sum_{u \sim v}(f(u) - f(v))^2}{\displaystyle\sum_v f(v)^2 d_v}$$

where P_i is the subspace generated by the harmonic eigenfunctions corresponding to λ_i, for $i \leq k-1$.

The different formulations for eigenvalues given above are useful in different settings and they will be used in later chapters. Here are some examples of special graphs and their eigenvalues.

EXAMPLE 1.1. For the complete graph K_n on n vertices, the eigenvalues are 0 and $n/(n-1)$ (with multiplicity $n-1$).

EXAMPLE 1.2. For the complete bipartite graph $K_{m,n}$ on $m+n$ vertices, the eigenvalues are 0, 1 (with multiplicity $m + n - 2$), and 2.

EXAMPLE 1.3. For the star S_n on n vertices, the eigenvalues are $0, 1$ (with multiplicity $n-2$), and 2.

EXAMPLE 1.4. For the path P_n on n vertices, the eigenvalues are $1 - \cos\frac{\pi k}{n-1}$ for $k = 0, 1, \cdots, n-1$.

EXAMPLE 1.5. For the cycle C_n on n vertices, the eigenvalues are $1 - \cos\frac{2\pi k}{n}$ for $k = 0, \cdots, n-1$.

EXAMPLE 1.6. For the n-cube Q_n on 2^n vertices, the eigenvalues are $\frac{2k}{n}$ (with multiplicity $\binom{n}{k}$) for $k = 0, \cdots, n$.

More examples can be found in Chapter 6 on explicit constructions.

1.3. Basic facts about the spectrum of a graph

Roughly speaking, half of the main problems of spectral theory lie in deriving bounds on the distributions of eigenvalues. The other half concern the impact and consequences of the eigenvalue bounds as well as their applications. In this section, we start with a few basic facts about eigenvalues. Some simple upper bounds and lower bounds are stated. For example, we will see that the eigenvalues of any graph lie between 0 and 2. The problem of narrowing the range of the eigenvalues for special classes of graphs offers an open-ended challenge. Numerous questions can be asked either in terms of other graph invariants or under further assumptions imposed on the graphs. Some of these will be discussed in subsequent chapters.

LEMMA 1.7. *For a graph G on n vertices, we have*

(i):

$$\sum_i \lambda_i \leq n$$

with equality holding if and only if G has no isolated vertices.

(ii): *For $n \geq 2$,*

$$\lambda_1 \leq \frac{n}{n-1}$$

with equality holding if and only if G is the complete graph on n vertices. Also, for a graph G without isolated vertices, we have

$$\lambda_{n-1} \geq \frac{n}{n-1}.$$

(iii): *For a graph which is not a complete graph, we have $\lambda_1 \leq 1$.*

(iv): *If G is connected, then $\lambda_1 > 0$. If $\lambda_i = 0$ and $\lambda_{i+1} \neq 0$, then G has exactly $i+1$ connected components.*

(v): *For all $i \leq n - 1$, we have*

$$\lambda_i \leq 2.$$

with $\lambda_{n-1} = 2$ if and only if a connected component of G is bipartite and nontrivial.

(vi): *The spectrum of a graph is the union of the spectra of its connected components.*

PROOF. (i) follows from considering the trace of $\mathcal{L}$.

The inequalities in (ii) follow from (i) and $\lambda_0 = 0$.

Suppose G contains two nonadjacent vertices a and b, and consider

$$f_1(v) = \begin{cases} d_b & \text{if } v = a, \\ -d_a & \text{if } v = b, \\ 0 & \text{if } v \neq a, b. \end{cases}$$

(iii) then follows from (1.2).

If G is connected, the eigenvalue 0 has multiplicity 1 since any harmonic eigenfunction with eigenvalue 0 assumes the same value at each vertex. Thus, (iv) follows from the fact that the union of two disjoint graphs has as its spectrum the union of the spectra of the original graphs.

(v) follows from equation (1.6) and the fact that

$$(f(x) - f(y))^2 \leq 2(f^2(x) + f^2(y)).$$

Therefore

$$\lambda_i \leq \sup_f \frac{\sum_{x \sim y}(f(x) - f(y))^2}{\sum_x f^2(x)d_x} \leq 2.$$

Equality holds for $i = n - 1$ when $f(x) = -f(y)$ for every edge $\{x, y\}$ in G. Therefore, since $f \neq 0$, G has a bipartite connected component. On the other hand, if G has a connected component which is bipartite, we can choose the function f so as to make $\lambda_{n-1} = 2$.

(vi) follows from the definition.   □

For bipartite graphs, the following slightly stronger result holds:

LEMMA 1.8. *The following statements are equivalent:*

(i): *G is bipartite.*

(ii): *G has $i+1$ connected components and $\lambda_{n-j} = 2$ for $1 \leq j \leq i$.*

(iii): *For each λ_i, the value $2 - \lambda_i$ is also an eigenvalue of G.*

PROOF. It suffices to consider a connected graph. Suppose G is bipartite graph with vertex set consisting of two parts A and B. For any harmonic eigenfunction f with eigenvalue λ, we consider the function g

$$g(x) = \left\{ \begin{array}{ll} f(x) & \text{if } x \in A, \\ -f(x) & \text{if } x \in B. \end{array} \right.$$

It is easy to check that g is a harmonic eigenfunction with eigenvalue $2 - \lambda$.   □

For a connected graph, we can immediately improve the lower bound of λ_1 in Lemma 1.7. For two vertices u and v, the *distance* between u and v is the number of edges in a shortest path joining u and v. The *diameter* of a graph is the maximum distance between any two vertices of G. Here we will give a simple eigenvalue lower bound in terms of the diameter of a graph. More discussion on the relationship between eigenvalues and diameter will be given in Chapter 3.

LEMMA 1.9. *For a connected graph G with diameter D, we have*

$$\lambda_1 \geq \frac{1}{D \text{ vol } G}$$

PROOF. Suppose f is a harmonic eigenfunction achieving λ_1 in (1.2). Let v_0 denote a vertex with $|f(v_0)| = \max_v |f(v)|$. Since $\sum_v f(v) = 0$, there exists a vertex u_0 satisfying $f(u_0)f(v_0) < 0$. Let P denote a shortest path in G joining u_0 and v_0. Then by (1.2) we have

$$
\begin{aligned}
\lambda_1 &= \frac{\displaystyle\sum_{x \sim y} (f(x) - f(y))^2}{\displaystyle\sum_x f^2(x) d_x} \\[2ex]
&\geq \frac{\displaystyle\sum_{\{x,y\} \in P} (f(x) - f(y))^2}{\text{vol } G \; f^2(v_0)} \\[2ex]
&\geq \frac{\frac{1}{D} (f(v_0) - f(u_0))^2}{\text{vol } G \; f^2(v_0)} \\[2ex]
&\geq \frac{1}{D \text{ vol } G}
\end{aligned}
$$

by using the Cauchy-Schwarz inequality.   □

LEMMA 1.10. *Let f denote a harmonic eigenfunction achieving λ_G in (1.2). Then, for any vertex $x \in V$, we have*

$$\frac{1}{d_x} \sum_{\substack{y \\ y \sim x}} (f(x) - f(y)) = \lambda_G f(x).$$

PROOF. We use a variational argument. For a fixed $x_0 \in V$, we consider f_ϵ such that

$$f_\epsilon(y) = \begin{cases} f(x_0) + \dfrac{\epsilon}{d_{x_0}} & \text{if } y = x_0, \\ f(y) - \dfrac{\epsilon}{\text{vol } G - d_{x_0}} & \text{otherwise.} \end{cases}$$

We have

$$\frac{\displaystyle\sum_{\substack{x,y \in V \\ x \sim y}} (f_\epsilon(x) - f_\epsilon(y))^2}{\displaystyle\sum_{x \in V} f_\epsilon^2(x) d_x}$$

$$= \frac{\displaystyle\sum_{\substack{x,y \in V \\ x \sim y}} (f(x) - f(y))^2 + \sum_{\substack{y \\ y \sim x_0}} \frac{2\epsilon(f(x_0) - f(y))}{d_{x_0}} - \sum_{\substack{y \\ y \neq x_0}} \sum_{\substack{y' \\ y \sim y'}} \frac{2\epsilon(f(y) - f(y'))}{\text{vol } G - d_{x_0}}}{\displaystyle\sum_{x \in V} f^2(x) d_x + 2\epsilon f(x_0) - \frac{2\epsilon}{\text{vol } G - d_{x_0}} \sum_{y \neq x_0} f(y) d_y}$$

$$+ O(\epsilon^2)$$

$$= \frac{\displaystyle\sum_{\substack{x,y \in V \\ x \sim y}} (f(x) - f(y))^2 + \frac{2\epsilon \displaystyle\sum_{\substack{y \\ y \sim x_0}} (f(x_0) - f(y))}{d_{x_0}} + \frac{2\epsilon \displaystyle\sum_{\substack{y \\ y \sim x_0}} (f(x_0) - f(y))}{\text{vol } G - d_{x_0}}}{\displaystyle\sum_{x \in V} f^2(x) d_x + 2\epsilon f(x_0) + \frac{2\epsilon f(x_0) d_{x_0}}{\text{vol } G - d_{x_0}}}$$

$$+ O(\epsilon^2)$$

since $\displaystyle\sum_{x \in V} f(x) d_x = 0$, and $\displaystyle\sum_y \sum_{y'} (f(y) - f(y')) = 0$. The definition in (1.2) implies that

$$\frac{\displaystyle\sum_{\substack{x,y \in V \\ x \sim y}} (f_\epsilon(x) - f_\epsilon(y))^2}{\displaystyle\sum_{x \in V} f_\epsilon^2(x) d_x} \geq \frac{\displaystyle\sum_{\substack{x,y \in V \\ x \sim y}} (f(x) - f(y))^2}{\displaystyle\sum_{x \in V} f^2(x) d_x}$$

If we consider what happens to the Rayleigh quotient for f_ϵ as $\epsilon \to 0$ from above, or from below, we can conclude that

$$\frac{1}{d_{x_0}} \sum_{\substack{y \\ y \sim x_0}} (f(x_0) - f(y)) = \lambda_G f(x_0).$$

and the Lemma is proved. $\qquad\square$

One can also prove the statement in Lemma 1.10 by recalling that $f = T^{-1/2}g$, where $\mathcal{L}g = \lambda_G g$. Then

$$T^{-1}Lf = T^{-1}(T^{1/2}\mathcal{L}T^{1/2})(T^{-1/2}g) = T^{-1/2}\lambda_G g = \lambda_G f,$$

and examining the entries gives the desired result.

With a little linear algebra, we can improve the bounds on eigenvalues in terms of the degrees of the vertices.

We consider the trace of $(I - \mathcal{L})^2$. We have

$$
\begin{aligned}
Tr(I - \mathcal{L})^2 &= \sum_i (1 - \lambda_i)^2 \\
&\leq 1 + (n-1)\bar{\lambda}^2,
\end{aligned}
$$

(1.9)

where

$$\bar{\lambda} = \max_{i \neq 0} |1 - \lambda_i|.$$

On the other hand,

(1.10)
$$
\begin{aligned}
Tr(I - \mathcal{L})^2 &= Tr(T^{-1/2}AT^{-1}AT^{-1/2}) \\
&= \sum_{x,y} \frac{1}{\sqrt{d_x}}A(x,y)\frac{1}{d_y}A(y,x)\frac{1}{\sqrt{d_x}} \\
&= \sum_x \frac{1}{d_x} - \sum_{x \sim y}(\frac{1}{d_x} - \frac{1}{d_y})^2,
\end{aligned}
$$

where A is the adjacency matrix. From this, we immediately deduce

LEMMA 1.11. *For a k-regular graph G on n vertices, we have*

(1.11)
$$\max_{i \neq 0} |1 - \lambda_i| \geq \sqrt{\frac{n - k}{(n-1)k}}$$

This follows from the fact that

$$\max_{i \neq 0} |1 - \lambda_i|^2 \geq \frac{1}{n-1}(tr(I - \mathcal{L})^2 - 1).$$

Let d_H denote the harmonic mean of the d_v's, i.e.,

$$\frac{1}{d_H} = \frac{1}{n}\sum_v \frac{1}{d_v}.$$

It is tempting to consider generalizing (1.11) with k replaced by d_H. This, however, is not true as shown by the following example due to Elizabeth Wilmer.

EXAMPLE 1.12. Consider the m-petal graph on $n = 2m+1$ vertices, $v_0, v_1, \cdots, v_{2m}$ with edges $\{v_0, v_i\}$ and $\{v_{2i-1}, v_{2i}\}$, for $i \geq 1$. This graph has eigenvalues $0, 1/2$ (with multiplicity $m - 1$), and $3/2$ (with multiplicity $m + 1$). So we have $\max_{i \neq 0}|1 - \lambda_i| = 1/2$. However,

$$\sqrt{\frac{n - d_H}{(n-1)d_H}} = \sqrt{\frac{m - 1/2}{2m}} \rightarrow \frac{1}{\sqrt{2}}$$

as $m \rightarrow \infty$.

Still, for a general graph, we can use the fact that

(1.12)
$$\frac{\sum_{x \sim y}(\frac{1}{d_x} - \frac{1}{d_y})^2}{\sum_{x \in V}(\frac{1}{d_x} - \frac{1}{d_H})^2 d_x} \le \lambda_{n-1} \le 1 + \bar{\lambda}.$$

Combining (1.9), (1.10) and (1.12), we obtain the following:

LEMMA 1.13. *For a graph G on n vertices, $\bar{\lambda} = \max_{i \neq 0}|1 - \lambda_i|$ satisfies*

$$1 + (n-1)\bar{\lambda}^2 \ge \frac{n}{d_H}(1 - (1+\bar{\lambda})(\frac{k}{d_H} - 1)),$$

where k denotes the average degree of G.

There are relatively easy ways to improve the upper bound for λ_1. From the characterization in the preceding section, we can choose any function $f : V(G) \to \mathbb{R}$, and its Rayleigh quotient will serve as an upper bound for λ_1. Here we describe an upper bound for λ_1 (see [205]).

LEMMA 1.14. *Let G be a graph with diameter $D \ge 4$, and let k denote the maximum degree of G. Then*

$$\lambda_1 \le 1 - 2\frac{\sqrt{k-1}}{k}\left(1 - \frac{2}{D}\right) + \frac{2}{D}.$$

One way to bound eigenvalues from above is to consider "contracting" the graph G into a weighted graph H (which will be defined in the next section). Then the eigenvalues of G can be upper-bounded by the eigenvalues of H or by various upper bounds on them, which might be easier to obtain. We remark that the proof of Lemma 1.14 proceeds by basically contracting the graph into a weighted path. We will prove Lemma 1.14 in the next section.

We note that Lemma 1.14 gives a proof (see [5]) that for any fixed k and for any infinite family of regular graphs with degree k,

$$\limsup \lambda_1 \le 1 - 2\frac{\sqrt{k-1}}{k}.$$

This bound is the best possible since it is sharp for the Ramanujan graphs (which will be discussed in Chapter 6). We note that the cleaner version of $\lambda_1 \le 1 - 2\sqrt{k-1}/k$ is not true for certain graphs (e.g., 4-cycles or complete bipartite graphs). This example also illustrates that the assumption in Lemma 1.14 concerning $D \ge 4$ is essential.

1.4. Eigenvalues of weighted graphs

Before defining weighted graphs, we will say a few words about two different approaches for giving definitions. We could have started from the very beginning with weighted graphs, from which simple graphs arise as a special case in which the weights are 0 or 1. However, the unique characteristics and special strength of graph theory is its ability to deal with the $\{0, 1\}$-problems arising in many natural situations. The clean formulation of a simple graph has conceptual advantages. Furthermore, as we shall see, all definitions and subsequent theorems for simple graphs can usually be easily carried out for weighted graphs. A weighted undirected graph G (possibly with loops) has associated with it a weight function $w : V \times V \to \mathbb{R}$ satisfying

$$w(u, v) = w(v, u)$$

and

$$w(u, v) \geq 0.$$

We note that if $\{u, v\} \notin E(G)$, then $w(u, v) = 0$. Unweighted graphs are just the special case where all the weights are 0 or 1.

In the present context, the degree d_v of a vertex v is defined to be:

$$d_v = \sum_u w(u, v),$$

$$\text{vol } G = \sum_v d_v.$$

We generalize the definitions of previous sections, so that

$$L(u, v) = \begin{cases} d_v - w(v, v) & \text{if } u = v, \\ -w(u, v) & \text{if } u \text{ and } v \text{ are adjacent,} \\ 0 & \text{otherwise.} \end{cases}$$

In particular, for a function $f : V \to \mathbb{R}$, we have

$$Lf(x) = \sum_{\substack{y \\ x \sim y}} (f(x) - f(y))w(x, y).$$

Let T denote the diagonal matrix with the (v, v)-th entry having value d_v. The Laplacian of G is defined to be

$$\mathcal{L} = T^{-1/2} L T^{-1/2}.$$

In other words, we have

$$\mathcal{L}(u, v) = \begin{cases} 1 - \dfrac{w(v, v)}{d_v} & \text{if } u = v, \text{ and } d_v \neq 0, \\ -\dfrac{w(u, v)}{\sqrt{d_u d_v}} & \text{if } u \text{ and } v \text{ are adjacent,} \\ 0 & \text{otherwise.} \end{cases}$$

We can still use the same characterizations for the eigenvalues of the generalized versions of $\mathcal{L}$. For example,

$$(1.13) \qquad \lambda_G := \lambda_1 \ = \ \inf_{g \perp T^{1/2}\mathbf{1}} \frac{\langle g, \, \mathcal{L}g \rangle}{\langle g, \, g \rangle}$$

$$= \ \inf_{\substack{f \\ \sum f(x)d_x = 0}} \frac{\sum_{x \in V} f(x)Lf(x)}{\sum_{x \in V} f^2(x)d_x}$$

$$= \ \inf_{\substack{f \\ \sum f(x)d_x = 0}} \frac{\sum_{x \sim y}(f(x) - f(y))^2 w(x,y)}{\sum_{x \in V} f^2(x)d_x}.$$

A contraction of a graph G is formed by identifying two distinct vertices, say u and v, into a single vertex v^*. The weights of edges incident to v^* are defined as follows:

$$\begin{aligned} w(x, v^*) \ &= \ w(x, u) + w(x, v), \\ w(v^*, v^*) \ &= \ w(u, u) + w(v, v) + 2w(u, v). \end{aligned}$$

LEMMA 1.15. *If H is formed by contractions from a graph G, then*

$$\lambda_G \leq \lambda_H$$

The proof follows from the fact that an eigenfunction which achieves λ_H for H can be lifted to a function defined on $V(G)$ such that all vertices in G that contract to the same vertex in H share the same value.

Now we return to Lemma 1.14.

SKETCHED PROOF OF LEMMA 1.14:
Let u and v denote two vertices that are at distance $D \geq 2t + 2$ in G. We contract G into a path H with $2t + 2$ edges, with vertices $x_0, x_1, \ldots x_t, z, y_t, \ldots, y_2, y_1, y_0$ such that vertices at distance i from u, $0 \leq i \leq t$, are contracted to x_i, and vertices at distance j from v, $0 \leq j \leq t$, are contracted to y_j. The remaining vertices are contracted to z. To establish an upper bound for λ_1, it is enough to choose a suitable function f, defined as follows:

$$\begin{aligned} f(x_i) \ &= \ a(k-1)^{-i/2}, \\ f(y_j) \ &= \ b(k-1)^{-j/2}, \\ f(z) \ &= \ 0, \end{aligned}$$

where the constants a and b are chosen so that

$$\sum_x f(x)d_x = 0.$$

It can be checked that the Rayleigh quotient satisfies

$$\frac{\sum_{u \sim v}(f(u) - f(v))^2 w(u,v)}{\sum_v f(v)^2 d_v} \leq 1 - \frac{2\sqrt{k-1}}{k}\left(1 - \frac{1}{t+1}\right) + \frac{1}{t+1},$$

since the ratio is maximized when $w(x_i, x_{i+1}) = k(k-1)^{i-1} = w(y_i, y_{i+1})$. This completes the proof of the lemma. $\square$

1.5. Eigenvalues and random walks

In a graph G, a walk is just a sequence of vertices $(v_0, v_1, \cdots, v_s)$ with $\{v_{i-1}, v_i\} \in E(G)$ for all $1 \leq i \leq s$. A random walk is determined by the transition probabilities $P(u,v) = Prob(x_{i+1} = v | x_i = u)$, which are independent of i. Clearly, for each vertex u,

$$\sum_v P(u,v) = 1.$$

For any initial distribution $f : V \to \mathbb{R}$ with $\sum_v f(v) = 1$, the distribution after k

steps is just fP^k (i.e., a matrix multiplication with f viewed as a row vector where P is the matrix of transition probabilities). The random walk is said to be *ergodic* if there is a unique *stationary distribution* $\pi(v)$ satisfying

$$\lim_{s \to \infty} fP^s(v) = \pi(v).$$

It is easy to see that necessary conditions for the ergodicity of P are (i) *irreducibility*, i.e., for any $u, v \in V$, there exists some s such that $P^s(u,v) > 0$ (ii) *aperiodicity*, i.e., g.c.d. $\{s : P^s(u,v) > 0\} = 1$. As it turns out, these are also sufficient conditions. A major problem of interest is to determine the number of steps s required for P^s to be *close* to its stationary distribution, given an arbitrary initial distribution.

We say a random walk is *reversible* if

$$\pi(u)P(u,v) = \pi(v)P(v,u).$$

An alternative description for a reversible random walk can be given by considering a weighted connected graph with edge weights satisfying

$$w(u,v) = w(v,u) = \pi(v)P(v,u)/c$$

where c can be any constant chosen for the purpose of simplifying the values. (For example, we can take c to be the average of $\pi(v)P(v,u)$ over all (v,u) with $P(v,u) \neq 0$, so that the values for $w(v,u)$ are either 0 or 1 for a simple graph.) The random walk on a weighted graph has as its transition probabilities

$$P(u,v) = \frac{w(u,v)}{d_u},$$

where $d_u = \sum_z w(u,z)$ is the (weighted) degree of u. The two conditions for ergodicity are equivalent to the conditions that the graph be (i) connected and (ii) non-bipartite. From Lemma 1.7, we see that (i) is equivalent to $\lambda_1 > 0$ and

(ii) implies $\lambda_{n-1} < 2$. As we will see later in (1.15), together (i) and (ii) deduce ergodicity.

We remind the reader that an unweighted graph has $w(u, v)$ equal to either 0 or 1. The usual random walk on an unweighted graph has transition probability $1/d_v$ of moving from a vertex v to any one of its neighbors. The transition matrix P then satisfies

$$P(u, v) = \begin{cases} 1/d_u & \text{if } u \text{ and } v \text{ are adjacent,} \\ 0 & \text{otherwise.} \end{cases}$$

In other words,

$$fP(v) = \sum_{\substack{u \\ u \sim v}} \frac{1}{d_u} f(u)$$

for any $f : V(G) \to \mathbb{R}$.

It is easy to check that

$$P = T^{-1}A = T^{-1/2}(I - \mathcal{L})T^{1/2},$$

where A is the adjacency matrix.

In a random walk with an associated weighted connected graph G, the transition matrix P satisfies

$$\mathbf{1}TP = \mathbf{1}T$$

where $\mathbf{1}$ is the vector with all coordinates 1. Therefore the stationary distribution is exactly $\pi = \mathbf{1}T/\text{vol } G$, We want to show that when k is large enough, for any initial distribution $f : V \to \mathbb{R}$, fP^k converges to the stationary distribution.

First we consider convergence in the L_2 (or Euclidean) norm. Suppose we write

$$fT^{-1/2} = \sum_i a_i \phi_i,$$

where ϕ_i denotes the orthonormal eigenfunction associated with λ_i.

Recall that $\phi_0 = \mathbf{1}T^{1/2}/\sqrt{\text{vol } G}$ and $\| \cdot \|$ denotes the L_2-norm, so

$$a_0 = \frac{\langle fT^{-1/2}, \mathbf{1}T^{1/2}\rangle}{\|\mathbf{1}T^{1/2}\|} = \frac{1}{\sqrt{\text{vol } G}}$$

since $\langle f, \mathbf{1}\rangle = 1$. We then have

$$
\begin{aligned}
\|fP^s - \pi\| &= \|fP^s - \mathbf{1}T/\text{vol } G\| \\
&= \|fP^s - a_0\phi_0 T^{1/2}\| \\
&= \|fT^{-1/2}(I - \mathcal{L})^s T^{1/2} - a_0\phi_0 T^{1/2}\| \\
&= \|\sum_{i \neq 0}(1 - \lambda_i)^s a_i \phi_i T^{1/2}\| \\
&\leq (1 - \lambda')^s \frac{\max_x \sqrt{d_x}}{\min_y \sqrt{d_y}} \\
&\leq e^{-s\lambda'} \frac{\max_x \sqrt{d_x}}{\min_y \sqrt{d_y}}
\end{aligned}
$$

(1.14)

where

$$\lambda' = \begin{cases} \lambda_1 & \text{if } 1 - \lambda_1 \geq \lambda_{n-1} - 1 \\ 2 - \lambda_{n-1} & \text{otherwise.} \end{cases}$$

So, after $s \geq 1/\lambda' \log(\max_x \sqrt{d_x}/\epsilon \min_y \sqrt{d_y})$ steps, the L_2 distance between fP^s and its stationary distribution is at most ϵ.

Although λ' occurs in the above upper bound for the distance between the stationary distribution and the s-step distribution, in fact, only λ_1 is crucial in the following sense. Note that λ' is either λ_1 or $2 - \lambda_{n-1}$. Suppose the latter holds, i.e., $\lambda_{n-1} - 1 \geq 1 - \lambda_1$. We can consider a modified random walk, called the lazy walk, on the graph G' formed by adding a loop of weight d_v to each vertex v. The new graph has Laplacian eigenvalues $\tilde{\lambda}_k = \lambda_k/2 \leq 1$, which follows from equation (1.13). Therefore,

$$1 - \tilde{\lambda}_1 \geq 1 - \tilde{\lambda}_{n-1} \geq 0,$$

and the convergence bound in L_2 distance in (1.14) for the modified random walk becomes

$$2/\lambda_1 \log(\frac{\max_x \sqrt{d_x}}{\epsilon \min_y \sqrt{d_y}}).$$

In general, suppose a weighted graph with edge weights $w(u, v)$ has eigenvalues λ_i with $\lambda_{n-1} - 1 \geq 1 - \lambda_1$. We can then modify the weights by choosing, for some constant c,

(1.15) $$w'(u, v) = \begin{cases} w(v, v) + cd_v & \text{if } u = v \\ w(u, v) & \text{otherwise.} \end{cases}$$

The resulting weighted graph has eigenvalues

$$\lambda'_k = \frac{\lambda_k}{1 + c} = \frac{2\lambda_k}{\lambda_{n-1} + \lambda_k}$$

where

$$c = \frac{\lambda_1 + \lambda_{n-1}}{2} - 1 \leq \frac{1}{2}.$$

Then we have

$$1 - \lambda'_1 = \lambda'_{n-1} - 1 = \frac{\lambda_{n-1} - \lambda_1}{\lambda_{n-1} + \lambda_1}.$$

Since $c \leq 1/2$ and we have $\lambda'_k \geq 2\lambda_k/(2 + \lambda_k) \geq 2\lambda_k/3$ for $\lambda_k \leq 1$. In particular we set

$$\lambda = \lambda'_1 = \frac{2\lambda_1}{\lambda_{n-1} + \lambda_1} \geq \frac{2}{3}\lambda_1.$$

Therefore the modified random walk corresponding to the weight function w' has an improved bound for the convergence rate in L_2 distance:

$$\frac{1}{\lambda} \log \frac{\max_x \sqrt{d_x}}{\epsilon \min_y \sqrt{d_y}}.$$

We remark that for many applications in sampling, the convergence in L_2 distance seems to be too weak since it does not require convergence at each vertex. There are several stronger notions of distance several of which we will mention.

A strong notion of convergence that is often used is measured by the relative pointwise distance (see [225]): After s steps, the *relative pointwise distance* (r.p.d.) of P to the stationary distribution $\pi(x)$ is given by

$$\Delta(s) = \max_{x,y} \frac{|P^s(y,x) - \pi(x)|}{\pi(x)}.$$

Let ψ_x denote the characteristic function of x defined by:

$$\psi_x(y) = \begin{cases} 1 & \text{if } y = x, \\ 0 & \text{otherwise.} \end{cases}$$

Suppose

$$\psi_x T^{1/2} = \sum_i \alpha_i \phi_i,$$

$$\psi_y T^{-1/2} = \sum_i \beta_i \phi_i.$$

where ϕ_i's denote the eigenfunction of the Laplacian $\mathcal{L}$ of the weighted graph associated with the random walk. In particular,

$$\alpha_0 = \frac{d_x}{\sqrt{\text{vol } G}},$$

$$\beta_0 = \frac{1}{\sqrt{\text{vol } G}}.$$

Let A^* denote the transpose of A. We have

$$
\begin{aligned}
\Delta(t) &= \max_{x,y} \frac{|\psi_y P^t \psi_x^* - \pi(x)|}{\pi(x)} \\
&= \max_{x,y} \frac{|\psi_y T^{-1/2} (I - \mathcal{L})^t T^{1/2} \psi_x^* - \pi(x)|}{\pi(x)} \\
&\leq \max_{x,y} \frac{\sum_{i \neq 0} |(1 - \lambda_i)^t \alpha_i \beta_i|}{d_x/\text{vol } G} \\
&\leq \bar{\lambda}^t \max_{x,y} \frac{\sum_{i \neq 0} |\alpha_i \beta_i|}{d_x/\text{vol } G} \\
&= \bar{\lambda}^t \max_{x,y} \frac{\|\psi_x T^{1/2}\| \, \|\psi_y T^{-1/2}\|}{d_x/\text{vol } G} \\
&\leq \bar{\lambda}^t \frac{\text{vol } G}{\min_{x,y} \sqrt{d_x d_y}} \\
&\leq e^{-t(1-\bar{\lambda})} \frac{\text{vol } G}{\min_x d_x}
\end{aligned}
$$

where $\bar{\lambda} = \max_{i \neq 0} |1 - \lambda_i|$. So if we choose t such that

$$t \geq \frac{1}{1 - \bar{\lambda}} \log \frac{\text{vol } G}{\epsilon \min_x d_x},$$

then, after t steps, we have $\Delta(t) \leq \epsilon$.

When $1 - \lambda_1 \neq \bar{\lambda}$, we can improve the above bound by using a lazy walk as described in (1.15). The proof is almost identical to the above calculation except for using the Laplacian of the modified weighted graph associated with the lazy walk. This can be summarized by the following theorem:

THEOREM 1.16. *For a weighted graph G, we can choose a modified random walk P so that the relative pairwise distance $\Delta(t)$ is bounded above by:*

$$\Delta(t) \leq e^{-t\lambda} \frac{\text{vol } G}{\min_x d_x} \leq \exp^{-2t\lambda_1/(2+\lambda_1)} \frac{\text{vol } G}{\min_x d_x}.$$

where $\lambda = \lambda_1$ if $2 \geq \lambda_{n-1} + \lambda_1$ and $\lambda = 2\lambda_1/(\lambda_{n-1} + \lambda_1)$ otherwise.

COROLLARY 1.17. *For a weighted graph G, we can choose a modified random walk P so that we have*

$$\Delta(t) \leq e^{-c}$$

if

$$t \geq \frac{1}{\lambda} \log \frac{\text{vol } G}{\min_x d_x}$$

where $\lambda = \lambda_1$ if $2 \geq \lambda_{n-1} + \lambda_1$ and $\lambda = 2\lambda_1/(\lambda_{n-1} + \lambda_1)$ otherwise.

We remark that for any initial distribution $f : V \to \mathbb{R}$ with $\langle f, 1 \rangle = 1$ and $f(x) \geq 0$, we have, for any x,

$$\frac{|f P^s(x) - \pi(x)|}{\pi(x)} \leq \sum_y f(y) \frac{|P^s(y, x) - \pi(x)|}{\pi(x)}$$

$$\leq \sum_y f(y) \Delta(s)$$

$$\leq \Delta(s).$$

Another notion of distance for measuring convergence is the so-called *total variation distance*, which is just half of the L_1 distance:

$$\Delta_{TV}(s) = \max_{A \subset V(G)} \max_{y \in V(G)} | \sum_{x \in A} (P^s(y, x) - \pi(x)) |$$

$$= \frac{1}{2} \max_{y \in V(G)} \sum_{x \in V(G)} | P^s(y, x) - \pi(x) |.$$

The total variation distance is bounded above by the relative pointwise distance, since

$$
\begin{aligned}
\Delta_{TV}(s) &= \max_{\substack{A \subset V(G) \\ \mathrm{vol} A \leq \frac{\mathrm{vol} G}{2}}} \max_{y \in V(G)} \mid \sum_{x \in A} (P^s(y,x) - \pi(x)) \mid \\
&\leq \max_{\substack{A \subset V(G) \\ \mathrm{vol} A \leq \frac{\mathrm{vol} G}{2}}} \sum_{x \in A} \pi(x) \Delta(s) \\
&\leq \frac{1}{2} \Delta(s).
\end{aligned}
$$

Therefore, any convergence bound using relative pointwise distance implies the same convergence bound using total variation distance. There is yet another notion of distance, sometimes called χ-*squared distance*, denoted by $\Delta'(s)$ and defined by:

$$
\begin{aligned}
\Delta'(s) &= \max_{y \in V(G)} \left(\sum_{x \in V(G)} \frac{(P^s(y,x) - \pi(x))^2}{\pi(x)} \right)^{1/2} \\
&\geq \max_{y \in V(G)} \sum_{x \in V(G)} \mid P^s(y,x) - \pi(x) \mid \\
&= 2\Delta_{TV}(s),
\end{aligned}
$$

using the Cauchy-Schwarz inequality. $\Delta'(s)$ is also dominated by the relative pointwise distance (which we will mainly use in this book).

$$
\begin{aligned}
\Delta'(s) &= \max_{x \in V(G)} \left(\sum_{y \in V(G)} \frac{(P^s(x,y) - \pi(y))^2}{\pi(y)} \right)^{1/2} \\
&\leq \max_{x \in V(G)} (\sum_{y \in V(G)} (\Delta(s))^2 \cdot \pi(y))^{\frac{1}{2}} \\
&\leq \Delta(s).
\end{aligned}
$$

We note that

$$
\begin{aligned}
\Delta'(s)^2 &\geq \sum_x \pi(x) \sum_y \frac{(P^s(x,y) - \pi(y))^2}{\pi(y)} \\
&= \sum_x \psi_x T^{1/2} (P^s - I_0) T^{-1} (P^s - I_0) T^{1/2} \psi_x^* \\
&= \sum_x \psi_x ((I - \mathcal{L})^{2s} - I_0) \psi_x^*,
\end{aligned}
$$

where I_0 denotes the projection onto the eigenfunction ϕ_0, ϕ_i denotes the i-th orthonormal eigenfunction of $\mathcal{L}$ and ψ_x denotes the characteristic function of x. Since

$$
\psi_x = \sum_i \phi_i(x) \phi_i,
$$

we have

$$(1.16) \qquad \Delta'(s)^2 \;\geq\; \sum_x \psi_x((I - \mathcal{L})^{2s} - I_0)\psi_x^*$$

$$= \sum_x (\sum_i \phi_i(x)\phi_i)((I - \mathcal{L})^{2s} - I_0)(\sum_i \phi_i(x)\phi_i)^*$$

$$= \sum_x \sum_{i \neq 0} \phi_i^2(x)(1 - \lambda_i)^{2s}$$

$$= \sum_{i \neq 0} \sum_x \phi_i^2(x)(1 - \lambda_i)^{2s}$$

$$= \sum_{i \neq 0}(1 - \lambda_i)^{2s}.$$

Equality in (1.16) holds if, for example, G is vertex-transitive, i.e., there is an automorphism mapping u to v for any two vertices in G, (for more discussions, see Chapter 7 on symmetrical graphs). Therefore, we conclude

THEOREM 1.18. *Suppose G is a vertex transitive graph. Then a random walk after s steps converges to the uniform distribution under total variation distance or χ-squared distance in a number of steps bounded by the sum of $(1 - \lambda_i)^{2s}$, where λ_i ranges over the non-trivial eigenvalues of the Laplacian:*

$$(1.17) \qquad \Delta_{TV}(s) \leq \frac{1}{2}\Delta'(s) = \frac{1}{2}(\sum_{i \neq 0}(1 - \lambda_i)^{2s})^{1/2}.$$

The above theorem is often derived from the *Plancherel formula*. Here we have employed a direct proof. We remark that for some graphs which are not vertex-transitive, a somewhat weaker version of (1.17) can still be used with additional work (see [81] and the remarks in Section 4.6). Here we will use Theorem 1.18 to consider random walks on an n-cube.

EXAMPLE 1.19. For the n-cube Q_n, our (lazy) random walk (as defined in (1.15)) converges to the uniform distribution under the total variation distance, as estimated as follows: From Example (1.6), the eigenvalues of the Q_n are $2k/n$ of multiplicity $\binom{n}{k}$ for $k = 0, \cdots, n$. The adjusted eigenvalues for the weighted graph corresponding to the lazy walk are $\lambda_k' = 2\lambda_k/(\lambda_{n-1} + \lambda_1) = \lambda_k n/(n + 1)$. By using Theorem 1.18 (also see [104]), we have

$$\Delta_{TV}(s) \leq \frac{1}{2}\Delta'(s) \;\leq\; \frac{1}{2}(\sum_{k=1}^n \binom{n}{k}(1 - \frac{2k}{n + 1})^{2s})^{1/2}$$

$$\leq \frac{1}{2}(\sum_{k=1}^n e^{k \log n - \frac{4ks}{n+1}})^{1/2}$$

$$\leq e^{-c}$$

if $s \geq \frac{1}{4}n \log n + cn$.

We can also compute the rate of convergence of the lazy walk under the relative pointwise distance. Suppose we denote vertices of Q_n by subsets of an n-set

$\{1, 2, \cdots, n\}$. The orthonormal eigenfunctions are ϕ_S for $S \subset \{1, 2, \cdots, n\}$ where

$$\phi_S(X) = \frac{(-1)^{|S \cap X|}}{2^{n/2}}$$

for any $X \subset \{1, 2, \cdots, n\}$. For a vertex indexed by the subset S, the characteristic function is denoted by

$$\psi_S(X) = \begin{cases} 1 & \text{if } X = S, \\ 0 & \text{otherwise.} \end{cases}$$

Clearly,

$$\psi_X = \sum_S \frac{(-1)^{|S \cap X|}}{2^{n/2}} \phi_S.$$

Therefore,

$$\begin{aligned} \frac{|P^s(X, Y) - \pi(Y)|}{\pi(Y)} &= |2^n \psi_X P^s \psi_Y^* - 1| \\ &\leq |2^n \psi_X P^s \psi_X^* - 1| \\ &= \sum_{S \neq \emptyset} (1 - \frac{2|S|}{n+1})^s \\ &= \sum_{k=1}^n \binom{n}{k} (1 - \frac{2k}{n+1})^s \end{aligned}$$

This implies

$$\begin{aligned} \Delta(s) &= \sum_{k=1}^n \binom{n}{k} (1 - \frac{2k}{n+1})^s \\ &\leq \sum_{k=1}^n e^{k \log n - \frac{2ks}{n+1}} \\ &\leq e^{-c} \end{aligned}$$

if

$$s \geq \frac{n \log n}{2} + cn.$$

So, the rate of convergence under relative pointwise distance is about twice that under the total variation distance for Q_n.

In general, $\Delta_{TV}(s), \Delta'(s)$ and $\Delta(s)$ can be quite different [81]. Nevertheless, a convergence lower bound for any of these notions of distance (and the L_2-norm) is λ^{-1}. This we will leave as an exercise. We remark that Aldous [4] has shown that if $\Delta_{TV}(s) \leq \epsilon$, then $P^s(y, x) \geq c_\epsilon \pi(x)$ for all vertices x, where c_ϵ depends only on ϵ.

Notes

For an induced subgraph of a graph, we can define the Laplacian with boundary conditions. We will leave the definitions for eigenvalues with Neumann boundary conditions and Dirichlet boundary conditions for Chapter 9.

The Laplacian for a directed graph is also very interesting. The Laplacian for a hypergraph has very rich structures. However, in this book we mainly focus on the Laplacian of a graph since the theory on these generalizations and extensions is still being developed.

In some cases, the factor $\log \frac{\text{vol } G}{\min_x d_x}$ in the upper bound for $\Delta(t)$ can be further reduced. Recently, P. Diaconis and L. Saloff-Coste [100] introduced a discrete version of the logarithmic Sobolev inequalities which can reduce this factor further for certain graphs (for $\Delta'(t)$). In Chapter 12, we will discuss some advanced techniques for further bounding the convergence rate under the relative pointwise distance.

CHAPTER 2

Isoperimetric problems

2.1. History

One of the earliest problems in geometry was the isoperimetric problem, which was considered by the ancient Greeks. The problem is to find, among all closed curves of a given length, the one which encloses the maximum area. The basic isoperimetric problem for graphs is essentially the same. Namely, remove as little of the graph as possible to separate out a subset of vertices of some desired "size". Here the size of a subset of vertices may mean the number of vertices, the number of edges, or some other appropriate measure defined on graphs. A typical case is to remove as few edges as possible to disconnect the graph into two parts of almost equal size. Such problems are usually called *separator* problems and are particularly useful in a number of areas including recursive algorithms, network design, and parallel architectures for computers, for example [**181**].

In a graph, a subset of edges which disconnects the graph is called a *cut*. Cuts arise naturally in the study of connectivity of graphs where the sizes of the disconnected parts are not of concern. Isoperimetric problems examine optimal relations between the size of the cut and the sizes of the separated parts. Many different names are used for various versions of isoperimetric problems (such as the conductance of a graph, the isoperimetric number, etc.) The concepts are all quite similar, but the differences are due to the varying definitions of cuts and sizes.

We distinguish two types of cuts. A *vertex-cut* is a subset of vertices whose removal disconnects the graph. Similarly, an *edge-cut* is a subset of edges whose removal separates the graph. The size of a subset of vertices depends on either the number of vertices or the number of edges. Therefore, there are several combinations.

Roughly speaking, isoperimetric problems involving edge-cuts correspond in a natural way to Cheeger constants in spectral geometry. The formulation and the proof techniques are very similar. Cheeger constants were studied in the thesis of Cheeger [**48**], but they can be traced back to Polyá and Szegö [**214**]. We will follow tradition and call the discrete versions by the same names, such as the Cheeger constant and the Cheeger inequalities.

2.2. The Cheeger constant of a graph

Before we discuss isoperimetric problems for graphs, let us first consider a measure on subsets of vertices. The typical measure just assigns weight 1 to each vertex, so the measure of a subset is its number of vertices. However, this implies that all vertices have the same measure. For some problems, this is appropriate only for regular graphs and does not work for general graphs. The measure we will use here takes into consideration the degree at a vertex. For a subset S of the vertices of G, we define vol S, the *volume* of S, to be the sum of the degrees of the vertices in S:

$$\text{vol } S = \sum_{x \in S} d_x,$$

for $S \subseteq V(G)$.

Next, we define the *edge boundary* ∂S of S to consist of all edges with exactly one endpoint in S:

$$\partial S = \{\{u, v\} \in E(G) : u \in S \text{ and } v \notin S\}$$

Let $\bar{S}$ denote the complement of S, i.e., $\bar{S} = V - S$. Clearly, $\partial S = \partial \bar{S} = E(S, \bar{S})$ where $E(A, B)$ denotes the set of edges with one endpoint in A and one endpoint in B. Similarly, we can define the *vertex boundary* δS of S to be the set of all vertices v not in S but adjacent to some vertex in S, i.e.,

$$\delta S = \{v \notin S : \{u, v\} \in E(G), u \in S\}.$$

We are ready to pose the following questions:

Problem 1: For a fixed number m, find a subset S with $m \le \text{vol } S \le \text{vol } \bar{S}$ such that the edge boundary ∂S contains as few edges as possible.

Problem 2: For a fixed number m, find a subset S with $m \le \text{vol } S \le \text{vol } \bar{S}$ such that the vertex boundary δS contains as few vertices as possible.

Cheeger constants are meant to answer exactly the questions above. For a subset $S \subset V$, we define

(2.1)
$$h_G(S) = \frac{|E(S, \bar{S})|}{\min(\text{vol } S, \text{vol } \bar{S})}.$$

The *Cheeger constant* h_G of a graph G is defined to be

(2.2)
$$h_G = \min_S h_G(S).$$

In some sense, the problem of determining the Cheeger constant is equivalent to solving Problem 1, since

$$|\partial S| \ge h_G \text{ vol } (S).$$

We remark that G is connected if and only if $h_G > 0$. We will only consider connected graphs. In a similar manner, we define the analogue of (2.1) for "vertex expansion" (instead of "edge expansion"). For a subset $S \subseteq V$, we define

$$(2.3) \qquad\qquad g_G(S) = \frac{\text{vol } \delta(S)}{\min(\text{vol } S, \text{vol } \bar{S})}$$

and

$$(2.4) \qquad\qquad g_G = \min_S g_G(S).$$

For regular graphs, we have

$$g_G(S) = \frac{|\delta(S)|}{\min(|S|, |\bar{S}|)}.$$

We define for a graph G (not necessarily regular)

$$\bar{g}_G(S) = \frac{|\delta(S)|}{\min(|S|, |\bar{S}|)}$$

and

$$\bar{g}_G = \min_S \bar{g}_G(S).$$

We remark that $\bar{g}$ is the corresponding Cheeger constant when the measure for each vertex is taken to be 1. More general measures will be considered later in Section 2.5. We note that both g_G and $\bar{g}_G$ are concerned with the vertex expansion of a graph and are useful for many problems.

2.3. The edge expansion of a graph

In this section, we focus on the fundamental relations between eigenvalues and the Cheeger constants. We first derive a simple upper bound for the eigenvalue λ_1 in terms of the Cheeger constant of a connected graph.

LEMMA 2.1.

$$2h_G \geq \lambda_1.$$

PROOF. We choose f based on an optimum edge cut C which achieves h_G and separates the graph G into two parts, A and B:

$$f(v) = \begin{cases} \dfrac{1}{\text{vol } A} & \text{if } v \text{ is in } A, \\[2ex] -\dfrac{1}{\text{vol } B} & \text{if } v \text{ is in } B. \end{cases}$$

By substituting f into (1.2), we have the following:

$$\begin{aligned} \lambda_1 &\leq |C|(1/\text{vol } A + 1/\text{vol } B) \\ &\leq \frac{2|C|}{\min(\text{vol } A, \text{vol } B)} \\ &= 2h_G. \end{aligned}$$

$\square$

Now, we will proceed to give a relatively short proof of an inequality in the opposite direction, so that we will have altogether

$$2h_G \geq \lambda_1 > \frac{h_G^2}{2}.$$

This is the so-called *Cheeger inequality* which often provides an effective way for bounding the eigenvalues of the graph.

THEOREM 2.2. *For a connected graph G,*

$$\lambda_1 > \frac{h_G^2}{2}.$$

PROOF. We consider the harmonic eigenfunction f of $\mathcal{L}$ with eigenvalue λ_1. We order vertices of G according to f. That is, relabel the vertices so that $f(v_i) \leq f(v_{i+1})$, for $1 \leq i \leq n-1$. Without loss of generality, we may assume

$$\sum_{f(v)<0} d_v \geq \sum_{f(u)\geq 0} d_u$$

For each i, $1 \leq i \leq |V|$, we consider the cut

$$C_i = \{ \{v_j, v_k\} \in E(G) \;:\; 1 \leq j \leq i < k \leq n \}.$$

We define α by

$$\alpha = \min_{1\leq i\leq n} \frac{|C_i|}{\min(\sum_{j\leq i} d_j, \sum_{j>i} d_j)}.$$

It is clear that $\alpha \geq h_G$. We consider the set V_+ of vertices v satisfying $f(v) \geq 0$ and the set E_+ of edges $\{u, v\}$ in G with either u or v in V_+. We define

$$g(x) = \begin{cases} f(x) & \text{if } u \in V_+ \\ 0 & \text{otherwise.} \end{cases}$$

We now have

$$\lambda_1 = \frac{\displaystyle\sum_{v\in V_+} f(v) \sum_{\{u,v\}\in E_+} (f(v) - f(u))}{\displaystyle\sum_{v\in V_+} f^2(v)d_v}$$

$$> \frac{\displaystyle\sum_{\{u,v\}\in E_+} (g(u) - g(v))^2}{\displaystyle\sum_{v\in V} g^2(v)d_v}$$

$$= \frac{\displaystyle\sum_{\{u,v\}\in E_+} (g(u) - g(v))^2 \sum_{\{u,v\}\in E_+} (g(u) + g(v))^2}{\displaystyle\sum_{v\in V} g^2(v)d_v \sum_{\{u,v\}\in E_+} (g(u) + g(v))^2}$$

$$\geq \frac{\left(\displaystyle\sum_{u\sim v} |g^2(u) - g^2(v)|\right)^2}{2\left(\displaystyle\sum_{v} g^2(v)d_v\right)^2}$$

$$\geq \frac{\left(\displaystyle\sum_{i} |g^2(v_i) - g^2(v_{i+1})| \, |C_i|\right)^2}{2\left(\displaystyle\sum_{v} g^2(v)d_v\right)^2}$$

$$\geq \frac{\left(\displaystyle\sum_{i} (g^2(v_i) - g^2(v_{i+1}))\alpha \sum_{j\leq i} d_j\right)^2}{2\left(\displaystyle\sum_{v} g^2(v)d_v\right)^2}$$

$$\geq \frac{\alpha^2}{2}$$

$$\geq \frac{h_G^2}{2}.$$

This completes the proof of Theorem 2.2.   □

We will state an improved version of Theorem 2.2 which however has a slightly more complicated proof.

THEOREM 2.3. *For any connected graph G, we always have*

$$\lambda_1 > 1 - \sqrt{1 - h_G^2}.$$

PROOF. From the proof of Theorem 2.2, we have

$$\lambda_1 \;=\; \frac{\displaystyle\sum_{v\in V_+} f(v) \sum_{u\sim v}(f(v)-f(u))}{\displaystyle\sum_{v\in V_+} f^2(v)d_v}$$

$$>\; \frac{\displaystyle\sum_{\{u,v\}\in E_+}(g(u)-g(v))^2}{\displaystyle\sum_{v\in V} g^2(v)d_v} \;=\; W.$$

Also, we have

$$W \;=\; \frac{(\displaystyle\sum_{\{u,v\}\in E_+}(g(u)-g(v))^2)\cdot(\displaystyle\sum_{\{u,v\}\in E_+}(g(u)+g(v))^2)}{(\displaystyle\sum_{v\in V} g^2(v)d_v)\cdot(\displaystyle\sum_{\{u,v\}\in E_+}(g(u)+g(v))^2)}$$

$$\geq\; \frac{(\displaystyle\sum_{u\sim v}|g^2(u)-g^2(v)|)^2}{(\displaystyle\sum_v g^2(v)d_v)\cdot(2\displaystyle\sum_v g^2(v)d_v - W\displaystyle\sum_v g^2(v)d_v)}$$

$$\geq\; \frac{(\displaystyle\sum_i |g^2(v_i)-g^2(v_{i+1})|\,|C_i|)^2}{(2-W)(\displaystyle\sum_v g^2(v))^2 d_v}$$

$$\geq\; \frac{(\displaystyle\sum_i (g^2(v_i)-g^2(v_{i+1}))\alpha\displaystyle\sum_{j\leq i} d_j)^2}{(2-W)(\displaystyle\sum_v g^2(v))^2 d_v}$$

$$\geq\; \frac{\alpha^2}{2-W}.$$

This implies that

$$W^2 - 2W + \alpha^2 \leq 0.$$

Therefore we have

$$\lambda_1 \;>\; W \;\geq\; 1-\sqrt{1-\alpha^2}$$
$$\geq\; 1-\sqrt{1-h_G^2}.$$

$\square$

We give a self-contained statement which has already been established in the above proof:

COROLLARY 2.4. *In a graph G with eigenfunction f associated with λ_1, we define, for each v,*

$$C_v = \{\,\{u,u'\}\in E(G)\;:\;f(u)\leq f(v)<f(u')\}$$

and

$$\alpha = \min_{v} \frac{|C_v|}{\min(\displaystyle\sum_{\substack{u \\ f(u) \leq f(v)}} d_u, \displaystyle\sum_{\substack{u \\ f(u) > f(v)}} d_u)}.$$

Then,

$$\lambda_1 > 1 - \sqrt{1 - \alpha^2}.$$

One immediate consequence is an improvement on the range of λ_1. For any connected (simple) graph G, we have

$$h_G \geq \frac{2}{\text{vol } G}.$$

Using Cheeger's inequality, we have

$$\lambda_1 > \frac{1}{2}(\frac{2}{\text{vol } G})^2 \geq \frac{2}{n^4}.$$

This lower bound is somewhat weaker than that in Lemma 1.9.

EXAMPLE 2.5. For a path P_n, the Cheeger constant is $1/\lfloor (n-1)/2 \rfloor$. As shown in Example 1.4, the eigenvalue λ_1 of P_n is $1 - \cos\frac{\pi}{n-1} \approx \frac{\pi^2}{2(n-1)^2}$. This shows that the Cheeger inequality in Theorem 2.2 is best possible up to within a constant factor.

EXAMPLE 2.6. For an n-cube Q_n, the Cheeger constant is $2/n$ which is equal to λ_1 (see Example 1.6). Therefore the inequality in Lemma 2.1 is sharp to within a constant factor.

Jerrum and Sinclair [167, 229] first used Cheeger's inequality as a main tool in deriving polynomial approximation algorithms for enumerating permanents and for other counting problems. The reader is referred to [228] for the related computational aspects of the Cheeger inequality.

2.4. The vertex expansion of a graph

The proofs of upper and lower bounds for the modified Cheeger constant g_G associated with vertex expansion are more complicated than those for edge expansion. This is perhaps due to the fact that the definition of h_G is in a way more natural and better scaled. Nevertheless, vertex expansion comes up often in many settings and it is certainly interesting in its own right.

Since $g_G \geq h_G$, we have

$$2g_G \geq \lambda_1.$$

For a general graph G, the eigenvalue λ_1 can sometimes be much smaller than $g_G^2/2$. One such example is given by joining two complete subgraphs by a matching. Suppose n is the total number of vertices. The eigenvalue λ_1 is no more than $8/n^2$, but g_G is large.

Still, it is desirable to have a lower bound for λ_1 in terms of g_G. Here we give a proof which is adapted from the argument first given by Alon [**5**].

THEOREM 2.7. *For a connected graph G,*

$$\lambda_1 > \frac{g_G^2}{4d + 2dg_G^2},$$

where d denotes the maximum degree of G.

PROOF. We follow the definition in the proof of Theorem 2.2. We have

$$
\lambda_1 = \frac{\displaystyle\sum_{v \in V_+} \sum_{u \sim v} (f(v) - f(u))f(v)}{\displaystyle\sum_{v \in V_+} d_v f^2(v)}
$$

$$
= \frac{\displaystyle\sum_{\substack{u \sim v \\ u,v \in V_+}} (f(v) - f(u))^2 + \sum_{\substack{u \sim v, v \in V_+ \\ u \notin V_+}} f(v)(f(v) - f(u))}{\displaystyle\sum_{v \in V_+} d_v f^2(v)}
$$

$$
> \frac{\displaystyle\sum_{u \sim v} (g(u) - g(v))^2}{\displaystyle\sum_v g^2(v) d_v},
$$

Now we use the max-flow min-cut theorem [**127**] as follows. Consider the network with vertex set $\{s, t\} \cup X \cup Y$ where s is the source, t is the sink, $X = V_+$ and Y is a copy of $V(G)$. The directed edges and their capacities are given as follows:

- For every u in X, the directed edge (s, u) has capacity $(1 + g_G)d_u$.
- For every $u \in X, v \in Y$, there is a directed edge (u, v) with capacity d_v if $\{u, v\} \in E$ or u is labelled by the same vertex as v in G.
- For every $v \in Y$, the directed edge (v, t) has large capacity, say, vol G.

To check that this network has its min-cut of size $(1 + g_G)$vol V_+, let C denote a cut separating s and t. Clearly C only contains edges from s and edges from X to Y. Let $X_1 = \{x \in X : \{s, x\} \notin C\}$. Then C separates X_1 from Y. Therefore the total capacity of the cut C is at least the sum of capacities of the edges $\{s, x\}, s \in X - X_1$, and the edges (u, v), $u \in X_1$ and $v \in X_1 \cup \delta X_1$. Since vol $(X_1 \cup \delta X_1) \geq (1 + g_G)$vol X_1, the total capacity of the cut is at least

$$(1 + g_G)\text{vol}(V_+ - X_1) + (1 + g_G)\text{vol } X_1 = (1 + g_G)\text{vol } V_+.$$

Since there is a cut of size $(1 + g_G)$vol V_+, we have proved that the min-cut is of size equal to $(1 + g_G)$vol V_+. By the max-flow min-cut theorem, there exists a flow function $F(u, v)$ for all directed edges in the network so that $F(u, v)$ is bounded above by the capacity of (u, v) and for each fixed $x \in X$ and $y \in Y$, we have

$$\sum_v F(x, v) = (1 + g_G)d_x,$$

$$\sum_v F(v, y) \leq d_y.$$

Then,

$$
\begin{aligned}
\sum_{\{u,v\} \in E} F^2(u, v)(f_+(u) + f_+(v))^2 &\leq 2 \sum_{\{u,v\} \in E} F^2(u, v)(f_+^2(u) + f_+^2(v)) \\
&= 2 \sum_v f_+^2(v) \Big(\sum_{\substack{u \\ \{u,v\} \in E}} F^2(u, v) \\
&\qquad + \sum_{\substack{v \\ \{u,v\} \in E}} F^2(v, u) \Big) \\
&\leq 2(1 + (1 + g_G^2)) \sum_u f_+^2(u) d_u^2 \\
&\leq 2d(2 + g_G^2) \sum_u f_+^2(u) d_u.
\end{aligned}
$$

Also,

$$
\begin{aligned}
\sum_{\{u,v\} \in E} F(u, v)(f_+^2(u) - f_+^2(v)) &= \sum_u f_+^2(u) \Big(\sum_{\substack{v \\ \{u,v\} \in E}} F(u, v) - \sum_{\substack{v \\ \{u,v\} \in E}} F(v, u) \Big) \\
&\geq g_G \sum_v f_+^2(v) d_v.
\end{aligned}
$$

Combining the above facts, we have

$$\lambda_1 \geq \frac{\displaystyle\sum_{\{u,v\}\in E}(f_+(u)-f_+(v))^2}{\displaystyle\sum_v f_+^2(v)d_v}$$

$$= \frac{\displaystyle\sum_{\{u,v\}\in E}(f_+(u)-f_+(v))^2 \sum_{\{u,v\}\in E}F^2(u,v)(f_+(u)+f_+(v))^2}{\displaystyle\sum_v f_+^2(v)d_v \sum_{\{u,v\}\in E}F^2(u,v)(f_+(u)+f_+(v))^2}$$

$$\geq \frac{\left(\displaystyle\sum_{\{u,v\}\in E}|F(u,v)(f_+^2(u)-f_+^2(v))|\right)^2}{\displaystyle\sum_v f_+^2(v)d_v\; 2d(2+g_G^2)\sum_v f_+^2(v)d_v}$$

$$\geq \frac{1}{4d+2dg_G^2}\left(\frac{\displaystyle\sum_{\{u,v\}\in E}F(u,v)(f_+^2(u)-f_+^2(v))}{\displaystyle\sum_v f_+^2(v)d_v}\right)^2$$

$$\geq \frac{g_G^2}{4d+2dg_G^2},$$

as desired.   □

EXAMPLE 2.8. For an n-cube, the vertex isoperimetric problem has been well studied. According to the Kruskal-Katona theorem [**179**, **171**], for a subset S of $\binom{n}{k}$ vertices, for $k \leq n/2$, the vertex boundary of S has at least $\binom{n}{k+1}$ vertices. Therefore, we have $g_{Q_n} = \binom{n}{n/2}2^{-(n-1)} \approx \sqrt{\frac{2}{\pi n}}$, for n even.

2.5. A characterization of the Cheeger constant

In this section, we consider a characterization of the Cheeger constant which has similar form to the Rayleigh quotient but with a different norm.

THEOREM 2.9. *The Cheeger constant h_G of a graph G satisfies*

(2.5)
$$h_G = \inf_f \sup_{c\in\mathbb{R}} \frac{\displaystyle\sum_{x\sim y}|f(x)-f(y)|}{\displaystyle\sum_{x\in V}|f(x)-c|d_x}$$

where f ranges over all functions $f : V \to \mathbb{R}$ which are not constant functions.

In language analogous to the continuous case, (2.5) can be thought of as

$$h_G = \inf_f \sup_{c\in\mathbb{R}} \frac{\displaystyle\int|\nabla f|}{\displaystyle\int|f-c|}.$$

PROOF. We choose c such that

$$\sum_{\substack{x \\ f(x)<c}} d_x \le \sum_{\substack{x \\ f(x)\ge c}} d_x$$

and

$$\sum_{\substack{x \\ f(x)\le c}} d_x > \sum_{\substack{x \\ f(x)>c}} d_x.$$

If $g = f - c$, then for $\sigma < 0$, we have

$$\sum_{\substack{x \\ g(x)<\sigma}} d_x \le \sum_{\substack{x \\ g(x)\ge\sigma}} d_x$$

and for $\sigma > 0$, we have

$$\sum_{\substack{x \\ g(x)<\sigma}} d_x \ge \sum_{\substack{x \\ g(x)>\sigma}} d_x.$$

We consider

$$\tilde{g}(\sigma) = |\{\{x,y\} \in E(G) : g(x) \le \sigma < g(y)\}|.$$

Then we have

$$
\begin{aligned}
\sum_{x \sim y} |f(x) - f(y)| &= \int_{-\infty}^{\infty} \tilde{g}(\sigma)d\sigma \\
&= \int_{-\infty}^{0} d\sigma \frac{\tilde{g}(\sigma)}{\displaystyle\sum_{g(x)<\sigma} d_x} \sum_{g(x)<\sigma} d_x + \int_{0}^{\infty} d\sigma \frac{\tilde{g}(\sigma)}{\displaystyle\sum_{g(x)>\sigma} d_x} \sum_{g(x)>\sigma} d_x \\
&\ge h_G \left(\int_{-\infty}^{0} d\sigma \sum_{g(x)<\sigma} d_x + \int_{0}^{\infty} d\sigma \sum_{g(x)>\sigma} d_x \right) \\
&= h_G \sum_{x \in V} |f(x) - c|d_x.
\end{aligned}
$$

In the opposite direction, suppose X is a subset of V satisfying

$$h_G = \frac{|E(X,\bar{X})|}{\text{vol } X}.$$

We consider a character function ψ defined by:

$$\psi(x) = \begin{cases} 1 & \text{if } x \in X \\ -1 & \text{otherwise.} \end{cases}$$

Then we have,

$$
\begin{aligned}
\sup_C \frac{\displaystyle\sum_{x \sim y} |\psi(x) - \psi(y)|}{\displaystyle\sum_{x \in V} |\psi(x) - C|d_x} &= \sup_C \frac{2|E(X,\bar{X})|}{(1-C)\text{vol}X + (1+C)\text{vol}\bar{X}} \\
&= \frac{2|E(X,\bar{X})|}{2\text{vol}X} \\
&= h_G.
\end{aligned}
$$

Therefore, we have

$$h_G \geq \inf_f \sup_{c \in \mathbb{R}} \frac{\displaystyle\sum_{x \sim y} |f(x) - f(y)|}{\displaystyle\sum_{x \in V} |f(x) - c| d_x}$$

and Theorem 2.9 is proved. $\square$

We will prove a variation of Theorem 2.9 which is not sharp but seems to be easier to use. Later on it will be used to derive an isoperimetric relationship between graphs and their cartesian products.

COROLLARY 2.10. *For a graph* G, *we have*

$$h_G \geq \inf_f \frac{\displaystyle\sum_{x \sim y} |f(x) - f(y)|}{\displaystyle\sum_{x \in V} |f(x)| d_x} \geq \frac{1}{2} h_G$$

where $f : V(G) \to \mathbb{R}$ *satisfies*

$$(2.6) \qquad\qquad \sum_{x \in V} f(x) d_x = 0.$$

PROOF. From Theorem 2.9, we already have

$$h_G \geq \inf_f \frac{\displaystyle\sum_{x \sim y} |f(x) - f(y)|}{\displaystyle\sum_{x \in V} |f(x)| d_x}$$

for f satisfying (2.6). It remains to prove the second part of the inequality. Suppose we define c as in the proof of Theorem 2.9. If $c \geq 0$, then we have

$$\sum_x |f(x)| d_x \quad \geq \quad \sum_{f(x) \geq 0} |f(x) - c| d_x$$

$$\geq \quad \frac{1}{2} \sum_x |f(x) - c| d_x.$$

If $c \leq 0$, then we have

$$\sum_x |f(x)| d_x \quad \geq \quad \sum_{f(x) \leq 0} |f(x) - c| d_x$$

$$\geq \quad \frac{1}{2} \sum_x |f(x) - c| d_x.$$

Therefore we have

$$\inf_f \frac{\displaystyle\sum_{x \sim y} |f(x) - f(y)|}{\displaystyle\sum_{x \in V} |f(x)| d_x} \geq \frac{1}{2} h_G.$$

The proof is complete. $\square$

Suppose we decide to have our measure be the number of vertices in S (and not the volume of S) for a subset S of vertices. We can then pose similar isoperimetric problems.

Problem 3: For a fixed number m, what is the minimum edge-boundary for a subset S of m vertices?

Problem 4: For a fixed number m, what is the minimum vertex-boundary for a subset S of m vertices?

We can define a modified Cheeger constant, which is sometimes called the *isoperimetric number*, by

$$h'(S) = \frac{|E(S, \bar{S})|}{\min(|S|, |\bar{S}|)}$$

and

$$h'_G = \inf_S h'(S).$$

We note that $h'_G \min_v d_v \leq h'_G \leq h'_G \max_v d_v$. These modified Cheeger constants are related to the eigenvalues of L, denoted by $0 = \lambda'_0 \leq \lambda'_1 \leq \ldots \leq \lambda'_{n-1}$, and

$$
\begin{aligned}
\lambda'_1 &= \inf_f \sup_c \frac{\sum_{u \sim v}(f(u) - f(v))^2}{\sum_v (f(v) - c)^2} \\
&= \inf \frac{\langle f, Lf \rangle}{\langle f, f \rangle}
\end{aligned}
$$

where f ranges over all functions f satisfying $\sum f(v) = 0$ which are not identical zero.

The above definition differs from that of $\mathcal{L}$ in (1.3) by the multiplicative factors of d_v for each term in the sum of the denominator. So, eigenvalues λ_i of $\mathcal{L}$ satisfy

$$0 \leq \lambda'_i \leq \lambda_i \max_v d_v.$$

By using methods similar to those in previous sections, we can show

$$2h'_G \geq \lambda'_1.$$

However, the lower bound for λ'_1 in terms of h'_G is a little messy in its derivation. We need to use the fact:

$$\sum_{u \sim v}(f(u) + f(v))^2 \leq 2\sum_v f(v)^2 d_v \leq 2\sum_v f(v)^2 \max_v d_v$$

in order to derive the modified Cheeger inequality:

$$\lambda'_1 \geq \frac{\lambda'^2_1}{2\max_v d_v}.$$

This is less elegant than the statement in Theorem 2.2.

We remark that the vertex expansion version of the Cheeger inequality are closely related to the so-called *expander graphs*, which we will examine further in Chapter 6.

2.6. Isoperimetric inequalities for cartesian products

Suppose G is a graph with a weight function w which assigns non-negative values to each vertex and each edge. A general Cheeger constant can be defined as follows:

$$h(G, w) = \min_S \frac{\displaystyle\sum_{\{x,y\} \in E(S,\bar{S})} w(x,y)}{\min(\displaystyle\sum_{x \in S} w(x), \sum_{y \notin S} w(y))}.$$

We say the weight function w is *consistent* if

$$\sum_u w(u,v) = w(v).$$

For example, the ordinary Cheeger constant is obtained by using the weight function $w_0(v) = d_v$ for any vertex v and $w_0(u,v) = 1$ for any edge $\{u,v\}$. Clearly, w_0 is consistent. On the other hand, the modified Cheeger constant is just $h'_G = h(G, w_1)$ where the weight function w_1 satisfies $w_1(u,v) = 1$ for any edge $\{u,v\}$ and $w_1(v) = 1$ for any vertex v. In this case, w_1 is not necessarily consistent. We note that graphs with consistent weight functions correspond in a natural way to random walks and reversible Markov chains. Namely, for a graph with a consistent weight function w, we can define the random walk with transition probability of moving from a vertex u to each of its neighbors v to be

$$P(u,v) = \frac{w(u,v)}{w(v)}.$$

Similar to Theorem 2.9, the general isoperimetric invariant $h(G, w)$ has the following characterization:

THEOREM 2.11. *For a graph G with weight function w, the isoperimetric invariant $h(G, w)$ of a graph G satisfies*

(2.7) $$h(G, w) = \inf_f \sup_{c \in \mathbb{R}} \frac{\displaystyle\sum_{x \sim y} |f(x) - f(y)| w(x,y)}{\displaystyle\sum_{x \in V} |f(x) - c| w(x)}$$

where f ranges over all $f : V \to \mathbb{R}$ which are not constant functions.

In particular, we also have the following characterization for the modified Cheeger constant.

THEOREM 2.12.

$$(2.8) \qquad h'_G = \inf_f \sup_{c \in \mathbb{R}} \frac{\displaystyle\sum_{x \sim y} |f(x) - f(y)|}{\displaystyle\sum_{x \in V} |f(x) - c|}$$

where f ranges over all $f : V \to \mathbb{R}$ which are not constant functions.

For two graphs G and H, the *cartesian product* $G \square H$ has vertex set $V(G) \times V(H)$ with (u, v) adjacent to (u', v') if and only if $u = u'$ and v is adjacent to v' in H, or $v = v'$ and u is adjacent to u' in G. For example, the cartesian product of n copies of one single edge is an n-cube, which is sometimes called a *hypercube*. The isoperimetric problem for n-cubes is an old and well-known problem. Just as in the continuous case where the sets with minimum vertex boundary form spheres, in a hypercube the subsets of given size with minimum vertex-boundary are so-called "Hamming balls", which consist of all vertices within a certain distance [**23, 156, 157, 189**]. The isoperimetric problems for grids (which are cartesian products of paths) and tori (which are cartesian products of cycles) have been well-studied in many papers [**33, 34, 249**].

We also consider a cartesian product of weighted graphs with consistent weight functions. For two weighted graphs G and G', with weight functions w, w', respectively, the weighted cartesian product $G \otimes G'$ has vertex set $V(G) \times V(G')$ with weight function $w \otimes w'$ defined as follows: For an edge $\{u, v\}$ in $E(G)$, we define $w \otimes w'((u, v'), (v, v')) = w(u, v)w'(v')$ and for an edge $\{u', v'\}$ in $E(G')$, we define $w \otimes w'((u, u'), (u, v')) = w(u)w'(u', v')$. We require $w \otimes w'$ to be consistent. Clearly, for a vertex $x = (u, v)$ in $G \otimes G'$, the weight of x in $G \otimes G'$ is exactly $2w(u)w'(v)$.

In general, for graphs G_i with consistent weight functions w_i, $i = 1, \cdots, k$, the weighted cartesian product $G_1 \otimes \cdots \otimes G_k$ has vertex set $V(G) \otimes \cdots \otimes V(G_k)$ with a consistent weight function $w_1 \otimes \cdots \otimes w_k$ defined as follows: For an edge $\{u, v\}$ in $E(G_i)$, the edge joining $(v_1, \cdots, v_{i-1}, u, v_{i+1}, \cdots, v_k)$ and $(v_1, \cdots, v_{i-1}, v, v_{i+1}, \cdots, v_k)$ has weight $w_1(v_1) \cdots w_{i-1}(v_{i-1})w_i(u, v)w_{i+1}(v_{i+1}), \cdots, w_k(v_k)$. We remark that $G_1 \otimes G_2 \otimes G_3$ is different from $(G_1 \otimes G_2) \otimes G_3$ or $G_1 \otimes (G_2 \otimes G_3)$.

The weighted cartesian product of graphs corresponds naturally to the cartesian product of random walks on graphs. Suppose $G_1, \cdots, G_k$ are graphs with the vertex sets $V(G_i)$. Each G_i is associated with a random walk with transition probability P_i as defined as in Section 1.5. The cartesian product of the random walks can be defined as follows: At the vertex $(v_1, \cdots, v_k)$, first choose a random "direction" i, between 1 and k, each with probability $1/k$. Then move to the vertex $(v_1, \cdots, v_{i-1}, u_i, v_{i+1}, \cdots, v_k)$ according to P_i. In other words,

$$P((v_1, \cdots, v_{i-1}, v_i, v_{i+1}, \cdots, v_k), (v_1, \cdots, v_{i-1}, u_i, v_{i+1}, \cdots, v_k)) = \frac{1}{k}P(v_i, u_i).$$

We point out that the above two notions of the cartesian products are closely related. In particular,

$$c\lambda_{G \square H} \le \lambda_{G \otimes H} \le c^{-1}\lambda_{G \square H}$$

where

$$c = \frac{min \ (min \ deg \ G, min \ deg \ H)}{max \ (max \ deg \ G, max \ deg \ H)}.$$

Here $min \ deg$ and $max \ deg$ denote the minimum degree and the maximum degree, respectively. The random walk on $G_1 \square \cdots \square G_k$ has transition probability P' of moving from a vertex $(v_1, \cdots, v_k)$ to the vertex $(v_1, \cdots, v_{i-1}, u_i, v_{i+1}, \cdots, v_k)$ given by:

$$P'((v_1, \cdots, v_{i-1}, v_i, v_{i+1}, \cdots, u_k), (v_1, \cdots, v_{i-1}, u_i, v_{i+1}, \cdots, v_k)) = \frac{w(v_i, u_i)}{\sum_{1 \leq j \leq k} w(v_j)}.$$

For a graph G, the natural consistent weight function associated with G has edge weight 1 and vertex weight d_x for any vertex x. Then we have

THEOREM 2.13. *The eigenvalue of a weighted cartesian product of* $G_1, G_2, \ldots,$ G_k *satisfies*

$$\lambda_{G_1 \otimes G_2 \otimes \cdots \otimes G_k} = \frac{1}{k} \min(\lambda_{G_1}, \lambda_{G_2}, \cdots, \lambda_{G_k})$$

where λ_G *denotes the first eigenvalue* λ_1 *of the graph* G.

Here we will give a proof for the case $k = 2$. Namely, we will show that the eigenvalue of a weighted cartesian product of G and H satisfies

(2.9) $$\lambda_{G \otimes H} = \frac{1}{2} \min(\lambda_G, \lambda_H).$$

PROOF. Without loss of generality, we assume that

$$\lambda_G \leq \lambda_H.$$

It is easy to see that

$$h_{G \otimes H} \leq \frac{1}{2} h_G.$$

Suppose $f : V(G) \to \mathbb{R}$ is the harmonic eigenfunction achieving λ_G. We choose a function $f_0 : V(G) \times V(H) \to \mathbb{R}$ by setting

$$f_0(u, v) = f(u).$$

Clearly, $\lambda_{G \otimes H}$ is less than the Rayleigh quotient using f_0 whose value is exactly $\lambda_G/2$.

In the opposite direction, we consider the harmonic eigenfunction $g : V(G) \times V(H) \to \mathbb{R}$ achieving $\lambda_{G \otimes H}$. We denote, for $u \in V(G), v \in V(H)$,

$$g_u = \frac{\sum_v g(u,v)d_v}{\text{vol } H},$$

$$g_v = \frac{\sum_u g(u,v)d_u}{\text{vol } G},$$

(2.10)
$$c = \frac{\sum_{u,v} g(u,v)d_u d_v}{\text{vol } G \text{ vol } H}.$$

Here, we repeatedly use the definition of eigenvalues and the Cauchy-Schwarz inequality:

$$\lambda_{G \otimes H} = \frac{\sum_v \sum_{u \sim u'} (g(u,v) - g(u',v))^2 d_v + \sum_u \sum_{v \sim v'} (g(u,v) - g(u,v'))^2 d_u}{\sum_{u,v} (g(u,v) - c)^2 2 d_u d_v}$$

$$\geq \frac{\lambda_G \sum_{u,v} (g(u,v) - g_v)^2 d_u d_v + (\sum_u d_u) \sum_{v \sim v'} (g_v - g_{v'})^2}{\sum_{u,v} (g(u,v) - g_v)^2 2 d_u d_v + \sum_{u,v} (g_v - c)^2 2 d_u d_v}$$

$$\geq \frac{\lambda_G \sum_{u,v} (g(u,v) - g_v)^2 d_u d_v + \lambda_H (\sum_u d_u) \sum_{v \sim v'} (g_v - c)^2 d_v}{\sum_{u,v} (g(u,v) - g_v)^2 2 d_u d_v + \sum_{u,v} (g_v - c)^2 2 d_u d_v}$$

$$\geq \frac{\lambda_G}{2}.$$

This completes the proof of (2.9).   □

THEOREM 2.14. *The Cheeger constant of a weighted cartesian product of $G_1, G_2,$ $\cdots, G_k$ satisfies*

$$\frac{1}{k} \min(h(G_1), h(G_2), \cdots, h(G_k)) \geq h(G_1 \otimes G_2 \otimes \cdots \otimes G_k)$$

$$\geq \frac{1}{2k} \min(h(G_1), h(G_2), \cdots, h(G_k)).$$

Here we again will prove the case for the product of two graphs and leave the proof of the general case as an exercise.

(2.11) $$\frac{1}{2} \min(h_G, h_H) \geq h_{G \otimes H} \geq \frac{1}{4} \min(h(G), h(H)).$$

PROOF. Without loss of generality, we assume that

$$h_G \leq h_H.$$

First we note that

$$h_{G \otimes H} \leq \frac{h_G}{2}.$$

Suppose $f : V(G) \to \mathbb{R}$ is a function achieving $h(G)$ in (2.7). We choose a function $f_0 : V(G) \times V(H) \to \mathbb{R}$ by setting

$$f_0(u, v) = f(u).$$

Clearly, $h_{G \otimes H}$ is no more than the value for the quotient of (2.7) using f_0 whose value is exactly $h_G/2$.

It remains to show that $h_{G \otimes H} \geq h_G/4$. To this end, we will repeatedly use Corollary 2.10, and we adopt the notation in the proof of (2.9).

$$h_{G \otimes H} = \frac{\displaystyle\sum_v \sum_{u \sim u'} |g(u, v) - g(u', v)| d_v + \sum_u \sum_{v \sim v'} |g(u, v) - g(u, v')| d_u}{\displaystyle\sum_{u,v} |g(u, v) - c| \, 2 d_u d_v}$$

$$\geq \frac{h_G \displaystyle\sum_{u,v} |g(u, v) - g_v| d_u d_v + \left(\sum_u d_u\right) \sum_{v \sim v'} |g_v - g_{v'}|}{\displaystyle\sum_{u,v} |g(u, v) - g_v| \, 2 d_u d_v + \sum_{u,v} |g_v - c| \, 2 d_u d_v}$$

$$\geq \frac{\dfrac{h_G}{2} \displaystyle\sum_{u,v} |g(u, v) - g_v| d_u d_v + \dfrac{h_H}{2} \left(\sum_u d_u\right) \sum_{v \sim v'} |g_v - c| d_v}{\displaystyle\sum_{u,v} |g(u, v) - g_v| \, 2 d_u d_v + \sum_{u,v} |g_v - c| \, 2 d_u d_v}$$

$$\geq \frac{h_G}{4}.$$

This completes the proof of (2.11).   □

For the modified Cheeger constant h_G', a similar isoperimetric inequality can be obtained:

COROLLARY 2.15. *The modified Cheeger constant of the cartesian product of $G_1, G_2, \ldots, G_k$ satisfies*

$$\min(h_{G_1}', h_{G_2}', \cdots, h_{G_k}') \geq h_{G_1 \square G_2 \square \cdots \square G_k}'$$
$$\geq \frac{1}{2} \min(h_{G_1}', h_{G_2}', \cdots, h_{G_k}').$$

The proof is quite similar to that of (2.11) (also see [**57**]) and will be omitted.

Notes

The characterization of the Cheeger constant in Theorem 2.9 is basically the Rayleigh quotient using the L_1-norm both in the numerator and denominator. In general, we can consider the so-called *Sobolev constants* for all $p, q > 0$:

$$
s_{p,q} = \inf_f \frac{\left(\sum_{u \sim v} |f(u) - f(v)|^p \right)^{1/p}}{\left(\sum_v |f(v)|^q d_v \right)^{1/q}}
$$

$$
= \inf_f \frac{\|\nabla f\|_p}{\|f\|_q}
$$

where f ranges over functions satisfying

$$
\sum_x |f(x) - c|^q d_x \geq \sum_x |f(x)|^q d_x
$$

for any c, or, equivalently,

$$
\int |f - c|^q \geq \int |f|^q.
$$

The eigenvalue λ_1 is associated with the case of $p = q = 2$, while the Cheeger constant corresponds to the case of $p = q = 1$. Some of the general cases will be considered later in Chapter 11 on Sobolev inequalities.

This chapter is mainly based on [52]. More general cases of the cartesian products are discussed in [57]. Another reference for weighted Cheeger constants and related isoperimetric inequalities is [83]. For graphs with "large" eigenvalue gaps, a polynomial time algorithm for finding an edge-separator with size within a constant factor of the optimum can be found in [61].

CHAPTER 3

Diameters and eigenvalues

3.1. The diameter of a graph

In a graph G, the distance between two vertices u and v, denoted by $d(u, v)$, is defined to be the length of a shortest path joining u and v in G. (It is possible to define the distance by various more general measures). The diameter of G, denoted by $D(G)$, is the maximum distance over all pairs of vertices in G. The diameter is one of the key invariants in a graph which is not only of theoretical interest but also has a wide range of applications. When graphs are used as models for communication networks, the diameter corresponds to the delays in passing messages through the network, and therefore plays an important role in performance analysis and cost optimization.

Although the diameter is a combinatorial invariant, it is closely related to eigenvalues. This connection is based on the following simple observation:

Let M denote an $n \times n$ matrix with rows and columns indexed by the vertices of G. Suppose G satisfies the property that $M(u, v) = 0$ if u and v are not adjacent. Furthermore, suppose we can show that for some integer t, and some polynomial $p_t(x)$ of degree t, we have

$$p_t(M)(u, v) \neq 0$$

for all u and v. Then we can conclude that the diameter $D(G)$ satisfies:

$$D(G) \leq t.$$

Suppose we take M to be the sum of the adjacency matrix and the identity matrix and the polynomial $p_t(x)$ to be just $(1 + x)^t$. The following inequality for regular graphs which are not complete graphs can then be derived (which will be proved in Section 3.2 as a corollary to Theorem 3.1; also see [51]):

$$(3.1) \qquad D(G) \leq \left\lceil \frac{\log(n - 1)}{\log(1/(1 - \lambda))} \right\rceil.$$

Here, λ basically only depends on λ_1. For example, we can take $\lambda = \lambda_1$ if $1 - \lambda_1 \geq \lambda_{n-1} - 1$. In general, we can slightly improve (3.1) by using the same "spectrum shifting" trick as in Section 1.5 (see Section 3.2). Namely, we define $\lambda = 2\lambda_1/(\lambda_{n-1} + \lambda_1) \geq 2\lambda_1/(2 + \lambda_1)$, and we then have

$$(3.2) \qquad D(G) \leq \left\lceil \frac{\log(n - 1)}{\log \dfrac{\lambda_{n-1} + \lambda_1}{\lambda_{n-1} - \lambda_1}} \right\rceil.$$

We note that for some graphs the above bound gives a pretty good upper bound for the diameter. For example, for k-regular Ramanujan graphs (defined later in 6.3.6.), we have $1 - \lambda_1 = \lambda_{n-1} - 1 = 1/(2\sqrt{k-1})$ so we get $D \leq \log(n-1)/(2\log(k-1))$ which is within a factor of 2 of the best possible bound.

The bound in (3.1) can be further improved by choosing p_t to be the Chebyshev polynomial of degree t. We can then replace the logarithmic function by $\cosh^{-1}$ (see [53] and Theorem 3.3) :

$$D(G) \leq \left\lceil \frac{\cosh^{-1}(n-1)}{\cosh^{-1} \frac{\lambda_{n-1}+\lambda_1}{\lambda_{n-1}-\lambda_1}} \right\rceil .$$

The above inequalities can be generalized in several directions. Instead of considering distances between two vertices, we can relate the eigenvalue λ_1 to distances between two subsets of vertices (see Section 3.2). Furthermore, for any $k \geq 1$, we can relate the eigenvalue λ_k to distances among $k+1$ distinct subsets of vertices (see Section 3.3).

We will derive several versions of the diameter-eigenvalue inequalities. From these inequalities, we can deduce a number of isoperimetric inequalities which are closely related to expander graphs which will also be discussed in Chapter 6.

It is worth mentioning that the above discrete methods for bounding eigenvalues can be used to derive new eigenvalue upper bounds for compact smooth Riemannian manifolds [54, 55]. This will be discussed in the last section of this chapter.

In contrast to many other more complicated graph invariants, the diameter is easy to compute. The diameter is the least integer t such that the matrix $M = I + A$ has the property that all entries of M^t are nonzero. This can be determined by using $O(\log n)$ iterations of matrix multiplication. Using the current best known bound $\mathcal{M}(n)$ for matrix multiplication where

$$\mathcal{M}(n) = O(n^{2.376})$$

this diameter algorithm requires at most $O(\mathcal{M}(n)\log n)$ steps. The problem of determining distances of all pairs of vertices for an undirected graph can also be done in $O(\mathcal{M}(n)\log n)$ time. Seidel [226] gave a simple recursive algorithm by reducing this problem for a graph G to a graph G' in which $u \sim v$ if $d(u,v) \leq 2$. (For directed graphs, an $O(\sqrt{\mathcal{M}(n)}n^3)$ algorithm can be found in [10].)

Another related problem is to find shortest paths between all pairs of vertices, which can be easily done in $O(n^3)$ steps (in fact $O(nm)$ is enough for a graph on n vertices and m edges). Apparently, we cannot compute all shortest paths explicitly in $o(n^3)$ time since some graphs can have cn^2 pairs of vertices having shortest paths of length at least $c'n$ each. However, we can compute a data structure that allows all shortest paths be constructed in time proportional to their lengths. For example, a matrix has its (u,v)-entry to be a neighbor of u in a shortest path connecting u and v. Seidel gave a randomized algorithm [226] to compute such a matrix in expected time $O(\mathcal{M}(n)\log n)$.

3.2. Eigenvalues and distances between two subsets

For two subsets X, Y of vertices in G, the distance between X and Y, denoted by $d(X, Y)$, is the minimum distance between a vertex in X and a vertex in Y, i.e.,

$$d(X, Y) = \min\{d(x, y) : x \in X, y \in Y\}.$$

Let $\bar{X}$ denote the complement of X in $V(G)$.

THEOREM 3.1. *Suppose G is not a complete graph. For $X, Y \subset V(G)$ and $X \neq \bar{Y}$, we have*

$$(3.3) \qquad d(X, Y) \leq \left\lceil \frac{\log \sqrt{\frac{\operatorname{vol} \bar{X} \operatorname{vol} \bar{Y}}{\operatorname{vol} X \operatorname{vol} Y}}}{\log \frac{\lambda_{n-1} + \lambda_1}{\lambda_{n-1} - \lambda_1}} \right\rceil.$$

PROOF. For $X \subset V(G)$, we define

$$\psi_X(x) = \begin{cases} 1 & \text{if } x \in X, \\ 0 & \text{otherwise}. \end{cases}$$

If we can show that for some integer t and some polynomial $p_t(z)$ of degree t,

$$\langle T^{1/2}\psi_Y, p_t(\mathcal{L})(T^{1/2}\psi_X) \rangle > 0$$

then there is a path of length at most t joining a vertex in X to a vertex in Y. Therefore we have $d(X, Y) \leq t$.

Let a_i denote the Fourier coefficients of $T^{1/2}\psi_X$, i.e.,

$$T^{1/2}\psi_X = \sum_{i=0}^{n-1} a_i \phi_i,$$

where the ϕ_i's are orthogonal eigenfunctions of $\mathcal{L}$. In particular, we have

$$a_0 = \frac{\langle T^{1/2}\psi_X, T^{1/2}\mathbf{1} \rangle}{\langle T^{1/2}\mathbf{1}, T^{1/2}\mathbf{1} \rangle} = \frac{\operatorname{vol} X}{\operatorname{vol} G}.$$

Similarly, we write

$$T^{1/2}\psi_Y = \sum_{i=0}^{n-1} b_i \phi_i.$$

Suppose we choose $p_t(z) = (1 - \frac{2z}{\lambda_1 + \lambda_{n-1}})^t$. Since G is not a complete graph, $\lambda_1 \neq \lambda_{n-1}$, and

$$|p_t(\lambda_i)| \leq (1 - \lambda)^t$$

for all $i = 1, \cdots, n-1$, where $\lambda = 2\lambda_1/(\lambda_{n-1} + \lambda_1)$. Therefore, we have

$$
\begin{aligned}
\langle T^{1/2}\psi_Y, p_t(\mathcal{L})T^{1/2}\psi_X \rangle &= a_0 b_0 + \sum_{i>0} p_t(\lambda_i) a_i b_i \\
&\geq a_0 b_0 - (1-\lambda)^t \sqrt{\sum_{i>0} a_i^2 \sum_{i>0} b_i^2} \\
&= \frac{\text{vol } X \text{ vol } Y}{\text{vol } G} - (1-\lambda)^t \frac{\sqrt{\text{vol } X \text{ vol } \bar{X} \text{ vol } Y \text{ vol } \bar{Y}}}{\text{vol } G}.
\end{aligned}
$$

by using the fact that

$$
\begin{aligned}
\sum_{i>0} a_i^2 &= \|T^{1/2}\psi_X\|^2 - \frac{(\text{vol } X)^2}{\text{vol } G} \\
&= \frac{\text{vol } X \text{ vol } \bar{X}}{\text{vol } G}.
\end{aligned}
$$

We note that in the above inequality, the equality holds if and only if $a_i = cb_i$ for some constant c for all i. This can only happen when $X = Y$ or $X = \bar{Y}$. Since the theorem obviously holds for $X = Y$ and we have the hypothesis that $X \neq \bar{Y}$, we may assume that the inequality is strict. If we choose

$$
t \geq \frac{\log \sqrt{\frac{\text{vol } \bar{X} \text{ vol } \bar{Y}}{\text{vol } X \text{ vol } Y}}}{\log \frac{1}{1-\lambda}}
$$

we have

$$
\langle T^{1/2}\psi_Y, p_t(\mathcal{L})T^{1/2}\psi_X \rangle > 0.
$$

This completes the proof of Theorem 3.3. $\square$

As an immediate consequence of Theorem 3.3, we have

COROLLARY 3.2. *Suppose G is a regular graph which is not complete. Then*

$$
D(G) \leq \left\lceil \frac{\log(n-1)}{\log \frac{\lambda_{n-1}+\lambda_1}{\lambda_{n-1}-\lambda_1}} \right\rceil.
$$

To improve the inequality in Theorem 3.3 in some cases, we consider Chebyshev polynomials:

$$
\begin{aligned}
T_0(z) &= 1, \\
T_1(z) &= z, \\
T_{t+1}(z) &= 2zT_t(z) - T_{t-1}(z), \quad \text{for integer } t > 1.
\end{aligned}
$$

Equivalently, we have

$$
T_t(z) = \cosh(t \cosh^{-1}(z)).
$$

In place of $p_t(\mathcal{L})$, we will use $S_t(\mathcal{L})$, where

$$
S_t(x) = \frac{T_t(\frac{\lambda_1 + \lambda_{n-1} - 2x}{\lambda_{n-1} - \lambda_1})}{T_t(\frac{\lambda_{n-1}+\lambda_1}{\lambda_{n-1}-\lambda_1})}.
$$

Then we have

$$\max_{x \in [\lambda_1, \lambda_{n-1}]} S_t(\lambda_1) \geq \frac{1}{T_t\left(\frac{\lambda_{n-1}+\lambda_1}{\lambda_{n-1}-\lambda_1}\right)}.$$

Suppose we take

$$t \geq \frac{\cosh^{-1}\sqrt{\frac{\text{vol }\bar{X}\text{ vol }\bar{Y}}{\text{vol }X\text{ vol }Y}}}{\cosh^{-1}\frac{\lambda_{n-1}+\lambda_1}{\lambda_{n-1}-\lambda_1}}.$$

Then we have

$$\langle T^{1/2}\psi_Y, S_t(\mathcal{L})T^{1/2}\psi_X \rangle > 0.$$

THEOREM 3.3. *Suppose G is not a complete graph. For $X, Y \subset V(G)$ and $X \neq \bar{Y}$, we have*

$$d(X,Y) \leq \left\lceil \frac{\cosh^{-1}\sqrt{\frac{\text{vol}\bar{X}\text{ vol}\bar{Y}}{\text{vol}X\text{ vol}Y}}}{\cosh^{-1}\frac{\lambda_{n-1}+\lambda_1}{\lambda_{n-1}-\lambda_1}} \right\rceil.$$

As an immediate application of Theorem 3.3, we can derive a number of isoperimetric inequalities. For a subset $X \subset V$, we define the s-boundary of X by

$$\delta_s X = \{y : y \notin X \text{ and } d(x,y) \leq s, \text{ for some } x \in X\}.$$

Clearly, $\delta_1(x)$ is exactly the vertex boundary $\delta(x)$. Suppose we choose $Y = V - \delta_s X$ in (3.3). From the proof of Theorem 3.1, we have

$$0 = \langle T^{1/2}\psi_Y, (I-\mathcal{L})^t T^{1/2}\psi_X \rangle > \frac{\text{vol }X\text{vol }Y}{\text{vol }G} - (1-\lambda)^t\frac{\sqrt{\text{vol }X\text{vol }Y\text{vol }\bar{X}\text{vol }\bar{Y}}}{\text{vol }G}.$$

This implies

(3.4) $$(1-\lambda)^{2t}\text{vol }\bar{X}\text{vol }\bar{Y} \geq \text{vol }X\text{ vol }Y.$$

For the case of $t = 1$, we have the following.

LEMMA 3.4. *For all $X \subseteq V(G)$, we have*

$$\frac{\text{vol }\delta X}{\text{vol }X} \geq \frac{1 - (1-\lambda)^2}{(1-\lambda)^2 + \text{vol}X/\text{vol}\bar{X}}$$

where $\lambda = 2\lambda_1/(\lambda_{n-1} + \lambda_1)$.

PROOF. Lemma 3.4 clearly holds for complete graphs. Suppose G is not complete, and take $Y = \bar{X} - \delta X$ and $t = 1$. From the proof of Theorem 3.1, we have

$$\begin{aligned} 0 &= \langle T^{1/2}\psi_Y, p_t(\mathcal{L})T^{1/2}\psi_X \rangle \\ &> \frac{\text{vol }X\text{ vol }Y}{\text{vol }G} - (1-\lambda)\frac{\sqrt{\text{vol }X\text{ vol }\bar{X}\text{ vol }Y\text{ vol }\bar{Y}}}{\text{vol }G}. \end{aligned}$$

Thus

$$(1-\lambda)^2\text{vol }\bar{X}\text{ vol }\bar{Y} > \text{vol }X\text{ vol }Y.$$

Since $\bar{Y} = X \cup \delta X$, this implies

$$(1-\lambda)^2(\text{vol }G - \text{vol }X)(\text{vol }X + \text{vol }\delta X) > \text{vol }X\ (\text{vol }G - \text{vol }X - \text{vol }\delta X)$$

After cancellation, we obtain

$$\frac{\text{vol } \delta X}{\text{vol } X} \geq \frac{1 - (1 - \lambda)^2}{(1 - \lambda)^2 + \text{vol } X/\text{vol } \bar{X}}.$$

$\square$

COROLLARY 3.5. *For* $X \subseteq V(G)$ *with* vol $X \leq$ vol $\bar{X}$, *where* G *is not a complete graph, we have*

$$\frac{\text{vol } \delta X}{\text{vol } X} \geq \lambda$$

where $\lambda = 2\lambda_1/(\lambda_{n-1} + \lambda_1)$.

PROOF. This follows from the fact that

$$\frac{\text{vol } \delta X}{\text{vol } X} \geq \frac{1 - (1 - \lambda)^2}{1 + (1 - \lambda)^2} \geq \lambda$$

by using $\lambda \leq 1$. $\square$

For general t, by a similar argument, we have

LEMMA 3.6. *For* $X \subseteq V(G)$ *and any integer* $t > 0$,

$$\frac{\text{vol } \delta_t X}{\text{vol } X} \geq \frac{1 - (1 - \lambda)^{2t}}{(1 - \lambda)^{2t} + \text{vol} X/\text{vol} \bar{\bar{X}}}$$

where $\lambda = 2\lambda_1/(\lambda_{n-1} + \lambda_1)$.

LEMMA 3.7. *For an integer* $t > 0$ *and* $X \subseteq V(G)$ *with* vol $X \leq$ vol $\bar{X}$, *we have*

$$\frac{\text{vol } \delta_t X}{\text{vol } X} \geq \frac{1 - (1 - \lambda)^{2t}}{1 + (1 - \lambda)^{2t}}$$

where $\lambda = 2\lambda_1/(\lambda_{n-1} + \lambda_1)$.

Suppose we consider, for $X \subseteq V(G)$:

$$N_s^* X = X \cup \delta_s X.$$

As a consequence of Lemma 3.6, we have

LEMMA 3.8. *For* $X \subseteq V(G)$ *with* vol $X \leq$ vol $\bar{X}$ *and any integer* $t > 0$,

$$\frac{\text{vol } N_t^* X}{\text{vol } X} \geq \frac{1}{(1 - \lambda)^{2t} \frac{\text{vol} \bar{X}}{\text{vol} G} + \frac{\text{vol} X}{\text{vol} G}}.$$

We remark that the special case of Lemma 3.8 for a regular graph and $t = 1$ was first proved by Tanner [**236**] (also see [**9**]). This is the basic inequality for establishing the vertex expansion properties of a graph. We will return to this inequality in Chapter 6.

3.3. Eigenvalues and distances among many subsets

To generalize Theorem 3.1 to distances among k subsets of the vertices, we need the following geometric lemma [**54**].

LEMMA 3.9. *Let $x_1, x_2, ... x_{d+2}$ denote $d+2$ arbitrary vectors in d-dimensional Euclidean space. Then there are two of them, say, v_i, v_j ($i \neq j$) such that $\langle v_i, v_j \rangle \geq 0$.*

PROOF. We will prove this by induction. First, it is clearly true when $d = 1$. Assume that it is true for $(d - 1)$-dimensional Euclidean space for some $d > 1$. Suppose that each pair of the given vectors has a negative scalar product. Let P be a hyperplane orthogonal to x_{d+2} and let x_i' be the projection of x_i on P for $i = 1, 2, ... d + 1$. We claim that $\langle x_i', x_j' \rangle < 0$ provided $i \neq j$. Since $\langle x_i, x_{d+2} \rangle < 0$, for $i \leq d + 1$, all vectors x_i lie in the same half-space with respect to P, which implies that each of them can be represented in the form

$$x_i = x_i' + a_i e$$

where $a_i > 0$ and e is a unit vector orthogonal to P, and directed to the same half-space as all the x_i. Then we have

$$0 > \langle x_i, x_j \rangle = \langle x_i' - a_i e, x_j' - a_j e \rangle = \langle x_i', x_j' \rangle + a_i a_j$$

which implies $\langle x_i', x_j' \rangle < 0$. On the other hand, by the induction hypothesis, out of $d + 1$ vectors $x_i', i = 1, 2, ... d + 1$ in the $(d - 1)$-dimensional space P, there are two vectors with non-negative scalar product. This is a contradiction and the lemma is proved. □

THEOREM 3.10. *Suppose G is not a complete graph. For $X_i \subset V(G), i = 0, 1, \cdots, k$, we have*

$$\min_{i \neq j} d(X_i, X_j) \leq \max_{i \neq j} \left\lceil \frac{\log \sqrt{\frac{\text{vol}\bar{X}_i \, \text{vol}\bar{X}_j}{\text{vol}X_i \, \text{vol}X_j}}}{\log \frac{1}{1-\lambda_k}} \right\rceil$$

if $1 - \lambda_k \geq \lambda_{n-1} - 1$. and $X_i \neq \bar{X}_j$ for $i = 0, 1, \cdots, k$.

PROOF. Let X and Y denote two distinct subsets among the X_i's. We consider

$$\langle T^{1/2}\psi_Y, (I - \mathcal{L})^t T^{1/2}\psi_X \rangle \geq a_0 b_0 + \sum_{i=1}^{k-1}(1 - \lambda_i)^t a_i b_i - \sum_{i \geq k}(1 - \lambda_k)^t |a_i b_i|.$$

For each $X_i, i = 0, 1, \cdots, k$, we consider the vector consisting of the Fourier coefficients of the eigenfunctions $\varphi_1, \cdots, \varphi_{k-1}$ in the eigenfunction expansion of X_i. Suppose we define a scalar product for two such vectors $(a_1, \cdots, a_{k-1})$ and $(b_1, \cdots, b_{k-1})$ by

$$\sum_{i=1}^{k-1}(1 - \lambda_i)^t a_i b_i.$$

From Lemma 3.9, we know that we can choose two of the subsets, say, X and Y with their associated vectors satisfying

$$\sum_{i=1}^{k-1}(1-\lambda_i)^t a_i b_i \geq 0.$$

Therefore, we have

$$\langle T^{1/2}\psi_Y, (I-\mathcal{L})^t T^{1/2}\psi_X \rangle \; > \; \frac{\text{vol}X \, \text{vol}Y}{\text{vol}\,G} - (1-\lambda_k)^t \frac{\sqrt{\text{vol}X \, \text{vol}\bar{X} \, \text{vol}Y \, \text{vol}\bar{Y}}}{\text{vol}\,G}$$

and Theorem 3.10 is proved.   □

We note that the condition $1 - \lambda_k \geq \lambda_{n-1} - 1$ can be eliminated by modifying λ_k as in proof of Theorem 3.1:

THEOREM 3.11. *For* $X_i \subset V(G), i = 0, 1, \cdots, k$, *we have*

$$\min_{i \neq j} d(X_i, X_j) \leq \max_{i \neq j} \left\lceil \frac{\log \sqrt{\frac{\text{vol}\bar{X}_i \, \text{vol}\bar{X}_j}{\text{vol}X_i \, \text{vol}X_j}}}{\log \frac{\lambda_{n-1}+\lambda_k}{\lambda_{n-1}-\lambda_k}} \right\rceil$$

if $\lambda_k \neq \lambda_{n-1}$ *and* $X_i \neq \bar{X}_j$.

Another useful generalization of Theorem 3.10 is the following:

THEOREM 3.12. *For* $X_i \subset V(G), i = 0, 1, \cdots, k$, *we have*

$$\min_{i \neq j} d(X_i, X_j) \leq \min_{0 \leq j < k} \max_{i \neq j} \left\lceil \frac{\log \sqrt{\frac{\text{vol}\bar{X}_i \, \text{vol}\bar{X}_j}{\text{vol}X_i \, \text{vol}X_j}}}{\log \frac{\lambda_{n-j-1}+\lambda_{k-j}}{\lambda_{n-j-1}-\lambda_{k-j}}} \right\rceil$$

where j *satisfies* $\lambda_{k-j} \neq \lambda_{n-j-1}$ *and* $X_i \neq \bar{X}_j$.

PROOF. For each j, $1 \leq j \leq k-1$, we can use a very similar proof to that in Theorem 3.10, such that there are two of the subsets X and Y with their corresponding vectors satisfying

$$\sum_{i \in S}(1-\lambda_i)^t a_i b_i \geq 0$$

where $S = \{i : 1 \leq i \leq k-j \text{ or } n-j+1 \leq i \leq n-1\}$. The proof then follows.   □

3.4. Eigenvalue upper bounds for manifolds

There are many similarities between the Laplace operator on compact Riemannian manifolds and the Laplacian for finite graphs. While the Laplace operator for a manifold is generated by the Riemannian metric, for a graph it comes from the adjacency relation. Sometimes it is possible to treat both the continuous and discrete cases by a universal approach. The general setting is as follows:

1. an underlying space M with a finite measure μ;
2. a well-defined Laplace operator $\mathcal{L}$ on functions on M so that $\mathcal{L}$ is a self-adjoint operator in $L^2(M, \mu)$ with a discrete spectrum;

3. if M has a boundary then the boundary condition should be chosen so that it does not disrupt self-adjointness of $\mathcal{L}$;

4. a distance function $\text{dist}(x, y)$ on M so that $\mid \nabla \text{dist} \mid \leq 1$ for an appropriate notion of gradient.

For a finite connected graph (also denoted by M in this section), the metric μ can be defined to be the degree of each vertex. Together with the Laplacian $\mathcal{L}$, all the above properties are satisfied. In addition, we can consider an r-neighborhood of the support $supp_r f$ of a function f in $L^2(M, \mu)$ for $r \in \mathbb{R}$:

$$\text{supp}_r\, f = \{x \in M : \text{dist}(x, \text{supp}\, f) \leq r\}$$

where dist denotes the distance function in M. For a polynomial of degree s, denoted by p_s, then we have

(3.5) $\text{supp}\, p_s(\mathcal{L})f \subset \text{supp}_s f.$

Let M be a complete Riemannian manifold with finite volume and let $\mathcal{L}$ be the self-adjoint operator $-\Delta$, where Δ is the Laplace operator associated with the Riemannian metric on M (which will be defined later in (3.9), also see [**254**]). Or, we could consider a compact Riemannian manifold M with boundary and let $\mathcal{L}$ be a self-adjoint operator $-\Delta$ subject to the Neumann or Dirichlet boundary conditions (defined in (3.10)). We can still have the following analogous version of (3.5) for the s-neighborhood of the support of a function.

There exists a non-trivial family of bounded continuous functions $P_s(\lambda)$ defined on the spectrum $\text{Spec}\mathcal{L}$, where s ranges over $[0, +\infty)$, so that for any function $f \in L^2(M, \mu)$:

(3.6) $\text{supp} P_s(\mathcal{L})f \subset \text{supp}_s f.$

For example, we can choose $P_s(\lambda) = \cos(\sqrt{\lambda}s)$ which clearly satisfies the requirement in (3.6).

Let us define

$$p(s) = \sup_{\lambda \in \text{Spec}\mathcal{L}} \mid P_s(\lambda) \mid$$

and assume that $p(s)$ is locally integrable.

We consider

$$\Phi(\lambda) = \int_0^\infty \phi(s) P_s(\lambda) ds$$

where $\phi(s)$ be a measurable function on $(0, +\infty)$ such that

$$\int_0^\infty \mid \phi(s) \mid p(s) ds < \infty.$$

In particular, $\Phi(\lambda)$ is a bounded function on $\text{Spec}\mathcal{L}$, and we can apply the operator $\Phi(\mathcal{L})$ to any function in $L^2(M, \mu)$.

We will prove the following general lemma which will be useful later.

LEMMA 3.13. *If $f \in L^2(M, \mu)$ then*

$$\|\Phi(\mathcal{L})f\|_{L^2(M \setminus \mathrm{supp}_r f)} \leq \|f\|_2 \int_r^\infty |\phi(s)| p(s) ds$$

where $\|f\|_2 := \|f\|_{L^2(M, \mu)}$.

PROOF. Let us denote

$$w(x) = \Phi(\mathcal{L})f(x) = \int_0^\infty \phi(s) P_s(\mathcal{L}) f(x) ds.$$

If the point x is not in $\mathrm{supp}_r f$ then $P_s(\mathcal{L})f(x) = 0$ whenever $s \leq r$. Therefore, for those points

$$w(x) = \int_r^\infty \phi(s) P_s(\mathcal{L}) f(x) ds$$

and

$$
\begin{aligned}
\|w\|_{L^2(M \setminus \mathrm{supp}_r f)} &\leq \left\| \int_r^\infty \phi(s) P_s(\mathcal{L}) f(x) ds \right\|_2 \\
&\leq \int_r^\infty \|\phi(s) P_s(\mathcal{L}) f(x)\|_2 ds \\
&\leq \int_r^\infty |\phi(s)| p(s) \|f\|_2 ds.
\end{aligned}
$$

The proof is complete.   □

As an immediate consequence, we have

COROLLARY 3.14. *If $f, g \in L^2(M, \mu)$ and the distance between $\mathrm{supp} \, f$ and $\mathrm{supp} \, g$ is D, then*

$$(3.7) \qquad \left| \int_M f \Phi(\mathcal{L}) g \, d\mu \right| \leq \|f\|_2 \, \|g\|_2 \int_D^\infty |\phi(s)| p(s) ds$$

The integral on the left-hand side of (3.7) is reduced to one over the support of g which in turn is majorized by the integral over the exterior of $\mathrm{supp}_D f$. The rest of the proof follows by a straightforward application of the Cauchy-Schwarz inequality.

For the choice of $P_s(\lambda) = \cos(\sqrt{\lambda} s)$, suppose we select

$$\phi(s) = \frac{1}{\sqrt{\pi t}} e^{-\frac{s^2}{4t}}.$$

Then we have

$$\Phi(\lambda) = \int_0^\infty \phi(s) P_s(\lambda) ds = e^{-\lambda t}.$$

COROLLARY 3.15. *If $f, g \in L^2(M, \mu)$ and the distance between the supports of f and g is equal to D then*

$$(3.8) \qquad \left| \int_M f e^{-t\mathcal{L}} g \, d\mu \right| \leq \|f\|_2 \, \|g\|_2 \int_D^\infty \frac{1}{\sqrt{\pi t}} e^{-\frac{s^2}{4t}} ds.$$

Let us mention a similar but weaker inequality:

COROLLARY 3.16.

$$\left| \int_M f e^{-t\mathcal{L}} g \, d\mu \right| \leq \|f\|_2 \, \|g\|_2 e^{-\frac{D^2}{4t}}.$$

This inequality was proved in [93] [254] and is quite useful.

Let M be a smooth connected compact Riemannian manifold and Δ be a Laplace operator associated with the Riemannian metric, i.e., in coordinates x_1, $x_2, \cdots, x_n$,

$$(3.9) \qquad\qquad \Delta u = \frac{1}{\sqrt{g}} \sum_{i,j=1}^n \frac{\partial}{\partial x_i} \left(\sqrt{g} g^{ij} \frac{\partial u}{\partial x_j} \right)$$

where g^{ij} are the contravariant components of the metric tensor, $g = \det \|g_{ij}\|$, $g^{ij} = \|g_{ij}\|^{-1}$, and u is a smooth function on M.

If the manifold M has a boundary ∂M, we introduce a boundary condition

$$(3.10) \qquad\qquad \alpha u + \beta \frac{\partial u}{\partial \nu} = 0$$

where $\alpha(x), \beta(x)$ are non-negative smooth functions on M such that $\alpha(x) + \beta(x) > 0$ for all $x \in \partial M$.

For example, both Dirichlet and Neumann boundary conditions satisfy these assumptions.

The operator $\mathcal{L} = -\Delta$ is self-adjoint and has a discrete spectrum in $L^2(M, \mu)$, where μ denotes the Riemannian measure. Let the eigenvalues be denoted by $0 = \lambda_0 < \lambda_1 \leq \lambda_2 \leq \cdots$. Let $\text{dist}(x, y)$ be a distance function on $M \times M$ which is Lipschitz and satisfies

$$|\nabla \text{dist}(x, y)| \leq 1$$

for all $x, y \in M$. For example, $\text{dist}(x, y)$ may be taken to be the geodesic distance, but we don't necessarily assume this is the case.

We want to show the following (also see [54]):

THEOREM 3.17. *For two arbitrary measurable disjoint sets X and Y on M, we have*

$$(3.11) \qquad\qquad \lambda_1 \leq \frac{1}{dist(X,Y)^2} \left(1 + \log \frac{(\mu M)^2}{\mu X \mu Y} \right)^2.$$

Moreover, if we have $k + 1$ disjoint subsets $X_0, X_1, \cdots, X_k$ such that the distance between any pair of them is greater than or equal to $D > 0$, then we have for any $k \geq 1$,

$$(3.12) \qquad\qquad \lambda_k \leq \frac{1}{D^2} (1 + \sup_{i \neq j} \log \frac{(\mu M)^2}{\mu X_i \, \mu X_i})^2.$$

PROOF. Let us denote by ϕ_i the eigenfunction corresponding to the i-th eigenvalue λ_i and normalized in $L^2(M, \mu)$ so that $\{\phi_i\}$ is an orthonormal frame in $L^2(M, \mu)$. For example, if either the manifold has no boundary or the Dirichlet or Neumann boundary condition is satisfied, there is one eigenvalue 0 with the associated eigenfunction being the constant function:

$$\phi_0 = \frac{1}{\sqrt{\mu M}}.$$

The proof is based upon two fundamental facts about the heat kernel $p(x, y, t)$, which by definition is the unique fundamental solution to the heat equation

$$\frac{\partial}{\partial t}u(x, t) - \Delta u(x, t) = 0$$

with the boundary condition (3.10) if the boundary ∂M is non-empty. The first fact is the eigenfunction expansion

$$(3.13) \qquad p(x, y, t) = \sum_{i=0}^{\infty} e^{-\lambda_i t} \phi_i(x) \phi_i(y)$$

and the second is the following estimate (by using Corollary 3.16):

$$(3.14) \qquad \int_X \int_Y p(x, y, t) f(x) g(y) \mu(dx) \mu(dy) \leq \left(\int_X f^2 \int_Y g^2 \right)^{\frac{1}{2}} \exp\left(-\frac{D^2}{4t} \right)$$

for any functions $f, g \in L^2(M, \mu)$ and for any two disjoint Borel sets $X, Y \subset M$ where $D = \text{dist}(X, Y)$.

We first consider the case $k = 2$. We start by integrating the eigenvalue expansion (3.13) as follows:

(3.15)

$$I(f, g) \equiv \int_X \int_Y p(x, y, t) f(x) g(y) \mu(dx) \mu(dy) = \sum_{i=0}^{\infty} e^{-\lambda_i t} \int_X f\phi_i \int_Y g\phi_i.$$

We denote by f_i the Fourier coefficients of the function $f\psi_X$ with respect to the frame $\{\phi_i\}$ and by g_i those of $g\,\psi_Y$. Then

$$\begin{aligned} I(f, g) &= e^{-\lambda_0 t} f_0 g_0 + \sum_{i=1}^{\infty} e^{-\lambda_i t} f_i g_i \\ &\geq e^{-\lambda_0 t} f_0 g_0 - e^{-\lambda_1 t} \|f\psi_X\|_2 \, \|g\psi_Y\|_2 \end{aligned}$$

where we have used

$$\left| \sum_{i=1}^{\infty} e^{-\lambda_i t} f_i g_i \right| \leq e^{-\lambda_1 t} \left(\sum_{i=1}^{\infty} f_i^2 \sum_{i=1}^{\infty} g_i^2 \right)^{\frac{1}{2}} \leq e^{-\lambda_1 t} \|f\psi_X\|_2 \|g\psi_Y\|_2.$$

By comparing (3.16) and (3.14), we have

(3.16)

$$\exp(-\lambda_1) \|f\psi_X\|_2 \, \|g\psi_Y\|_2 \geq f_0 g_0 - \|f\psi_X\|_2 \, \|g\psi_Y\|_2 \exp\left(-\frac{D^2}{4t} \right).$$

We will choose t so that the second term on the right-hand side (3.16) is equal to one half of the first one (here we take advantage of the Gaussian exponential since it can be made arbitrarily close to 0 by taking t small enough):

$$t = \frac{D^2}{4 \log \frac{2\|f\psi_X\|_2 \, \|g\psi_Y\|_2}{f_0 g_0}}.$$

For this t we have

$$\exp(-\lambda_1)\|f\psi_X\|_2\|g\psi_Y\|_2 \geq \frac{1}{2}f_0g_0$$

which implies

$$\lambda_1 \leq \frac{1}{t}\log\frac{2\|f\psi_X\|_2\,\|g\psi_Y\|_2}{f_0g_0}.$$

After substituting this value of t, we have

$$\lambda_1 \leq \frac{4}{D^2}\left(\log\frac{2\|f\psi_X\|_2\,\|g\psi_Y\|_2}{f_0g_0}\right)^2.$$

Finally, we choose $f = g = \phi_0$ and take into account that

$$f_0 = \int_X f\phi_0 = \int_X \phi_0^2,$$

and

$$\|f\psi_X\|_2 = \left(\int_X \phi_0^2\right)^{\frac{1}{2}} = \sqrt{f_0}.$$

Similar identities hold for g. We then obtain

$$\lambda_1 \leq \frac{1}{D^2}\left(\log\frac{4}{\int_X \phi_0^2 \int_Y \phi_0^2}\right)^2.$$

Now we consider the general case $k > 2$. For a function $f(x)$, we denote by f_i^j the i-th Fourier coefficient of the function $f\mathbf{1}_{X_j}$ i.e.

$$f_i^j = \int_{X_j} f\phi_i.$$

Similar to the case of $k = 2$, we have

$$I_{lm}(f,f) = \int_{X_l}\int_{X_m} p(x,y,t)f(x)f(y)\mu(dx)\mu(dy).$$

Again, we have the following upper bound for $I_{lm}(f,f)$:

$$(3.17) \qquad I_{lm}(f,f) \leq \|f\psi_{X_l}\|_2\,\|f\psi_{X_m}\|_2\exp\left(-\frac{D^2}{4t}t\right).$$

We can rewrite the lower bound (3.16) in another way:

(3.18)

$$I_{lm}(f,f) \geq e^{-\lambda_1 t}f_0^l f_0^m + \sum_{i=1}^{k-1}e^{-\lambda_i t}f_i^l f_i^m - e^{-\lambda_k t}\|f\psi_{X_l}\|_2\,\|f\psi_{X_m}\|_2$$

Now we can eliminate the middle term on the right-hand side of (3.18) by choosing appropriate l and m. To this end, let us consider $k+1$ vectors $f^m = (f_1^m, f_2^m, ...f_{k-1}^m)$, $m = 0, 1, 2, ...k$ in $\mathbb{R}^{k-1}$ and let us endow this $(k-1)$-dimensional space with a scalar product given by

$$(v,w) = \sum_{i=1}^{k-1}v_i w_i e^{-\lambda_i t}.$$

By using Corollary 3.9, out of any $k + 1$ vectors in $(k-1)$-dimensional Euclidean space there are always two vectors with non-negative scalar product. So, we can find different l, m so that $(f^l, f^m) \geq 0$ and therefore we can eliminate the second term on the right-hand side (3.18).

Comparing (3.17) and (3.18), we have

(3.19)
$$e^{-\lambda_k t} \|f\psi_{X_l}\|_2 \, \|f\psi_{X_m}\|_2 \;\leq\; f_0^l f_0^m - \|f\psi_{X_l}\|_2 \, \|f\psi_{X_m}\|_2 \exp\left(-\frac{D^2}{4t}\right).$$

Similar to the case $k = 2$, we can choose t so that the right-hand side is at least $\frac{1}{2} f_0^l f_0^m$. We select

$$t = \min_{l \neq m} \frac{D^2}{4 \log \frac{2\|f\psi_{X_l}\|_2 \|f\psi_{X_m}\|_2}{f_0^l f_0^m}}.$$

From (3.19), we have

$$\lambda_k \leq \frac{1}{t} \log \frac{2\|f\psi_{X_l}\|_2 \, \|f\psi_{X_m}\|_2}{f_0^l f_0^m}.$$

By substituting t from above and taking $f = \phi_0$, (3.12) follows. $\qquad\square$

Although differential geometry and spectral graph theory share a great deal in common, there is no question that significant differences exist. Obviously, a graph is not "differentiable" and many geometrical techniques involving high-order derivatives could be very difficult, if not impossible, to utilize for graphs. There are substantial obstacles for deriving the discrete analogues of many of the known results in the continuous case. Nevertheless, there are many successful examples of developing the discrete parallels, and this process sometimes leads to improvement and strengthening of the original results from the continuous case. Furthermore, the discrete version often offers a different viewpoint which can provide additional insight to the fundamental nature of geometry. In particular, it is useful in focusing on essentials which are related to the global structure instead of the local conditions.

There are basically two approaches in the interplay of spectral graph theory and spectral geometry. One approach, as we have seen in this section, is to share the concepts and methods while the proofs for the continuous and discrete, respectively, remain self-contained and independent. The second approach is to approximate the discrete cases by continuous ones. This method is usually coupled with appropriate assumptions and estimates. One example of this approach will be given in Chapter 10.

For almost every known result in spectral geometry, a corresponding set of questions can be asked: Can the results be translated to graph theory? Is the discrete analogue true for graphs? Do the proof techniques still work for the discrete case? If not, how should the methods be modified? If the discrete analogue does not hold for general graphs, can it hold for some special classes of graphs? What are the characterizations of these graphs?

Discrete invariants are somewhat different from the continuous ones. For example, the number of vertices n is an important notion for a graph. Although it can be roughly identified as a quantity which goes to infinity in the continuous analog, it is of interest to distinguish $n, n \log n, n^2, \cdots$ and 2^n, for example. Therefore more careful analysis is often required. For Riemannian manifolds, the dimension of the manifold is usually given and can be regarded as a constant. This is however not true in general for graphs. The interaction between spectral graph theory and differential geometry opens up a whole range of interesting problems.

Notes

This chapter is based on the original diameter-eigenvalue bounds given in [51] and a subsequent paper [53]. The generalizations to pairs of subsets for regular graphs were given by Kahale in [169]. The generalizations to k subsets and to Riemannian manifolds can be found in [54, 55].

CHAPTER 4

Paths, flows, and routing

4.1. Paths and sets of paths

One of the main themes in graph theory concerns paths joining pairs of vertices. For example, the Hamiltonian path problem is to decide if a graph has a simple path containing every vertex of the graph. Some diameter and distance problems involve finding shortest paths. There are many basic problems depending on sets of paths that are either vertex-disjoint or edge-disjoint. These path problems arise naturally in a variety of guises, such as the study of communicating processes on networks, data flow on parallel computers, and the analysis of routing algorithms on VLSI chips. Some path problems appear to be quite difficult computationally. For example, the Hamiltonian path problem is well known to be NP-complete. The problem of finding disjoint paths between given pairs of vertices even in very special graphs [**140**] is also NP-complete. Nevertheless, we will see that eigenvalue techniques are amazingly effective in providing good solutions for a range of path problems.

Before we proceed, we first define several types of disjoint paths that we call *flow*, *route set*, and *routing*. Consider a graph G with vertex set V and edge set E. Suppose X and Y are two equinumerous subsets of vertices of G. In general, X and Y can be multisets and it is not necessary to require $X \cap Y = \emptyset$.

For $|X| = |Y| = m$, a *flow* F from X to Y consists of m paths in G joining the vertices in X to the vertices in Y. We call X the *input* of the flow F and Y the *output* of F. Paths in F join vertices of X to vertices in Y in a one-to-one fashion, but we do not care about "who is talking to whom." We *do* care that the paths be chosen so that no edge is overused. For example, the paths might be required to be edge-disjoint or vertex-disjoint or with small "congestion" in the sense that every edge (or vertex) of G is used in relatively few paths of F. We will define "congestion" precisely later.

A *route set* is a flow with input-output assignments. Namely, for a specified assignment $A = \{(x_i, y_i) : x_i \in X, y_i \in Y\}$, a route set consists of paths P_i joining x_i to y_i for each i. In other words, an assignment specifies "who is talking to whom."

Roughly speaking, a *routing* R is a dynamic version of a route set . It can be defined as a pebble game. Initially, there is a pebble p_i placed at each input vertex x_i with destination y_i for each of the assignments (x_i, y_i) in A. At each time unit, a pebble can be moved to some adjacent vertex. The routing R is then a route set together with a strategy for moving pebbles to their destinations. Additional requirements can be imposed. For example, at each time unit, the edges used for moving pebbles should be (vertex- or edge-) disjoint or all edges must have small congestion.

Flow and routes are very useful in establishing lower bounds for Cheeger constants as well as providing lower bounds for eigenvalues (see 4.2. and 4.5.). Conversely, for graphs with good eigenvalue lower bounds, short routes and effective routing schemes exist with small congestion which will be described in 4.3. and 4.4.

4.2. Flows and Cheeger constants

Flows are closely related to cuts as evidenced by the max flow-min cut theorem which was used in the previous section. In fact, there is a direct connection between the Cheeger constants and flow problems on graphs. Although these observations are quite easy, we will state them here since they are useful for bounding eigenvalues. We follow the definition for Cheeger constants h_G and h'_G as given in Section 2.2.

LEMMA 4.1. *For a graph G on n vertices, suppose there is a set of $\binom{n}{2}$ paths joining all pairs of vertices such that each edge of G is contained in at most m paths. Then*

$$h'_G = \sup_S \frac{|E(S, \bar{S})|}{\min(|S|, |\bar{S}|)} \geq \frac{n}{2m}$$

PROOF. The proof follows from the simple fact that for any set $S \subseteq V$ with $|S| \leq |\bar{S}|$, we have

$$\begin{aligned} |E(S, \bar{S})| \cdot m &\geq |S| \cdot |\bar{S}| \\ &\geq |S| \cdot \frac{n}{2}. \end{aligned}$$

$\square$

As an immediate consequence, we have the following:

COROLLARY 4.2. *For a k-regular graph G on n vertices, suppose there is a set P of $\binom{n}{2}$ paths joining all pairs of vertices such that each edge of G is contained in at most m paths in P. Then the Cheeger constant h_G satisfies*

$$h_G = \inf_S \frac{|E(S, \bar{S})|}{k \min(|S|, |\bar{S}|)} \geq \frac{n}{2mk}$$

By using Cheeger's inequality in Chapter 2 and the above lower bound for the Cheeger constant derived from a flow, we can establish eigenvalue lower bounds for a regular graph. In fact, we can derive a better lower bound for λ_1 directly from a flow in a general graph. We first prove a simple version for a regular graph.

THEOREM 4.3. *For a k-regular graph G on n vertices, suppose there is a set P of $\binom{n}{2}$ paths joining all pairs of vertices such that each path in P has length at most*

*l and each edge of G is contained in at most m paths in P. Then the eigenvalue λ_1
satisfies*

$$\lambda_1 \geq \frac{n}{kml}$$

PROOF. Using the definition (1.5) of the eigenvalues, we consider the harmonic
eigenfunction $f : V(G) \to \mathbb{R}$ achieving λ_1.

$$\lambda_1 \;=\; \frac{n \sum\limits_{\{x,y\}\in E(G)} (f(x) - f(y))^2}{k \sum\limits_{x,y} (f(x) - f(y))^2}$$

We note that for $x, y \in V(G)$ and the path $P(x, y)$ joining x and y in G, we
have

$$(f(x) - f(y))^2 \leq |P(x,y)| \sum_{e\in P(x,y)} f^2(e) \leq l \sum_{e\in P(x,y)} f^2(e),$$

where $f^2(e) = (f(x) - f(y))^2$ for $e = \{x, y\}$, and $|P(x, y)|$ denotes the number of
edges of G in $P(x, y)$. Hence

$$m \sum_{e\in E(G)} f^2(e) \;\geq\; \sum_{x,y} \sum_{e\in P(x,y)} f^2(e)$$

$$\geq\; \frac{1}{l} \sum_{x,y} (f(x) - f(y))^2.$$

Therefore we have

$$\lambda_1 \;\geq\; \frac{n}{kml}.$$

This completes the proof of Theorem 4.3 $\square$

For a general graph, the above theorems can be generalized as follows:

THEOREM 4.4. *For an undirected graph G, replace each edge $\{u, v\}$ by two
directed edges (u, v) and (v, u). Suppose there is a set P of $4e^2$ paths such that
for each (ordered) pair of directed edges there is a directed path joining them. In
addition, assume that each directed edge of G is contained in at most m directed
paths in P. Then the Cheeger constant h_G satisfies*

$$h_G = \min \frac{|E(S, \bar{S})|}{\min(\mathrm{vol}\, S, \mathrm{vol}\, \bar{S})} \geq \frac{\mathrm{vol}\, G}{2m}.$$

PROOF. For any $S \subseteq V(G)$, we have

$$m|E(S, \bar{S})| \geq \mathrm{vol}\, S \, \mathrm{vol}\, \bar{S} \geq \frac{\mathrm{vol}\, S \, \mathrm{vol}\, G}{2}.$$

$\square$

THEOREM 4.5. *For an undirected graph G, replace each edge $\{u, v\}$ by two directed edges (u, v) and (v, u). Suppose there is a set P of $4e^2$ directed paths such that for each (ordered) pair of directed edges there is a directed path joining them, each of length at most l. In addition, assume that each directed edge of G is contained in at most m directed paths in P. Then the eigenvalue λ_1 satisfies*

$$\lambda_1 \geq \frac{\text{vol } G}{ml}.$$

The proof of Theorem 4.5 is very similar to that of Theorem 4.3 and will be omitted.

We remark that Theorems 4.3, 4.5 can be generalized in a number of ways. For example, instead of having one path joining two vertices, we can ask for a number of paths or weighted paths with fixed total capacities (in the spirit of the max flow-min cut theorem). Another direction is to derive the comparison theorems which will be discussed in Section 4.5.

4.3. Eigenvalues and routes with small congestion

In a graph G, a random walk of length l starting at a vertex v of G is a randomly chosen sequence $v = v_0, v_1, \ldots v_l$, where each v_{i+1} is chosen, uniformly at random and independently, among the neighbors of v_i, for $i = 0, \ldots, l-1$. We say that the walk visits v_i at time i.

In a graph G with $\lambda_1 > 0$, a random walk starting from any vertex converges roughly in $\frac{\log n}{\lambda_1}$ steps to the stationary distribution (if G is bipartite, we use a lazy random walk; see Section 1.5.) We will use this property to derive the following fact.

THEOREM 4.6. *Let G be a graph on n vertices and suppose $l \geq \log n/\lambda_1$. Suppose for any $v \in V(G)$ there are d_v random walks of length l starting at v. For any edge q, let $I(q)$ denote the total number of walks containing q. Then, almost surely (i.e., with probability tending to 1 as n tends to infinity), there is no edge q so that*

$$I(q) > 10l.$$

PROOF. Let P denote the transition matrix defined by

$$P(u, v) = \begin{cases} 1/d_u & \text{if } u \text{ and } v \text{ are adjacent} \\ 0 & \text{otherwise.} \end{cases}$$

The probability that a random walk $W(u)$ starting at u visits a vertex x at time i is precisely $\psi_u P^i(\psi_x)^*$ where ψ_y is the unit vector having 1 in coordinate y and 0 in every other coordinate. For a directed edge (u, v), the probability that a random walk $W(x)$ visits u at time i and v at time $i + 1$ is

$$\psi_x P^i \psi_u^* / d_u.$$

With d_v walks starting at v, the sum of the probabilities that there exists a walk $W(x)$ that visits $q = \{u, v\}$ is

$$I(q) \leq \sum_{i=0}^{l-1} \sum_{x} d_x \psi_x P^i \psi_u^* / d_u$$

$$= \sum_{i=0}^{l-1} (\mathbf{1}T) P^i \psi_u^* / d_u$$

where $\mathbf{1}$ denotes the all 1's vector, T is the diagonal matrix with entries $T(v, v) = d_v$, and

$$(\mathbf{1}T) P^j = \mathbf{1}T.$$

Therefore

$$I(q) \leq \sum_{i=0}^{l-1} \mathbf{1}T \psi_u^* / d_u$$

$$= l.$$

Therefore, for each fixed v, the expectation of the random variable $I(e)$ is no more than l. We observe that this random variable is a sum of $|E(G)|$ independent indicator random variables (see, e.g., [12] Theorem A.12, page 237), and that for each fixed edge q, the probability that $I(q)$ exceeds, say $10l$, is at most

$$\left(\frac{e^9}{10^{10}}\right)^l \ll \frac{1}{n^3}.$$

Since there are at most n^2 edges, it follow that the probability that there is an edge with $I(q) > 10l$ is much smaller than $\frac{1}{n}$. This completes the proof of Theorem 4.6. $\square$

The above estimate can in fact be proved directly. Suppose there is a set of m independent events such that the probability of the i-th event is p_i. Furthermore, suppose that $\sum p_i \leq \gamma$. Then the probability that at least s events occur is bounded by

$$\sum_{\substack{S \subset \{1,...m\} \\ |S|=s}} \prod_{i \in S} p_i \leq \frac{1}{s!} \left(\sum p_i\right)^s$$

$$\leq \left(\frac{\gamma e}{s}\right)^s.$$

The proof of Theorem 4.6 follows by choosing $\gamma = l$ and $s = 10\gamma$.

THEOREM 4.7. *Let G denote a graph on n vertices. Let $A = \{(x_i, y_i) : x_i \in X, y_i \in Y\}$ denote any assignment such that each vertex v is in X with multiplicity d_v and in Y with multiplicity d_v. Then there are paths P_i joining x_i to y_i of length at most $\frac{2}{\lambda_1} \log n$ such that each edge of G is contained in at most $\frac{20}{\lambda_1} \log n$ paths P_j.*

PROOF. Let P_i denote a random walk of length $2l$ between x_i and y_i where $l \approx \frac{\log n}{\lambda_1}$. Using an argument of Valiant in his work on parallel routing [246] (also see [40]), we may assume that each walk consists of two random walks of length

l , one starting from x_i and the other from y_i. The reason for this is that by our eigenvalue condition, the distribution of the random walk of length l is close to its stationary distribution and hence one may view the walk P_i as being chosen by first choosing its middle point (according to the stationary distribution) and then choosing its two halves. The proof of Theorem 4.7 then follows from Theorem 4.6.   □

The above proofs for Theorem 4.6 and 4.7 were adapted from the following simpler version for regular graphs in [**7**]:

THEOREM 4.8. *Let G denote a k-regular graph on n vertices. Let π denote a permutation of the vertices of G Then there are paths P_x joining x to $\pi(x)$ of length at most $\frac{2}{\lambda_1} \log n$ such that each edge of G is contained in at most $\frac{20}{k\lambda_1} \log n$ paths P_y.*

4.4. Routing in graphs

In this section, we consider a simple (though fundamental) problem of the following type: Suppose we are given a connected graph G with vertex set V and edge set E. Initially, each vertex v of G is occupied by a unique marker or "pebble" p_v. To each pebble p_v is associated a destination vertex $\pi(v) \in V$, so that distinct pebbles have distinct destinations. Pebbles can be moved to different vertices of G according to the following basic procedure: At each step a disjoint collection of edges of G is selected and the pebbles at each edge's two endpoints are interchanged. Our goal is to move or "route" the pebbles to their respective destinations in a minimum number of steps.

We will imagine the steps occurring at discrete times, and we let $p_v(t) \in V$ denote the location of the pebble with initial position v at time $t = 0, 1, 2, \dots$. Thus, for any t, the set $\{p_v(t) : v \in V\}$ is just a permutation of V. We will denote our target permutation that takes v to $\pi(v)$, $v \in V$, by π. Define $rt(G, \pi)$ to be the minimum possible number of steps to achieve π. Finally, define $rt(G)$, the *routing number* of G, by

$$rt(G) = \max_{\pi} \ rt(G, \pi)$$

where π ranges over all destination permutations on G. (Sometimes we will also call π a routing assignment.)

In more algebraic terms, the problem is simply to determine for G the largest number of terms $\tau = (u_1 v_1)(u_2 v_2) \cdots (u_r v_r)$ ever required to represent any permutation in the symmetric group on $n = |V|$ symbols, where each permutation τ consists of a product of disjoint transpositions $(u_k v_k)$ with all pairs $\{u_k, v_k\}$ required to be edges of G.

To see that $rt(G)$ always exists, let us restrict our attention to some spanning subtree T of G. It is clear that if p has destination which is a leaf of T, then we can first route p to its destination u, and then complete the routing on $T \setminus \{u\}$ by induction.

An obvious lower bound on $rt(G)$ is the following:

$$rt(G) \geq D(G)$$

where $D(G)$ denotes the diameter of G.

For P_n a path on n vertices, our routing problem reduces to a well studied problem in parallel sorting networks, the so-called *odd-even transposition sort* (see [176] for a comprehensive survey). In this case, it can be shown that $rt(P_n) = n$. In fact, any permutation π on P_n can be sorted in n steps by labelling consecutive edges in P_n as $e_1, e_2, \cdots, e_{n-1}$ and only making interchanges with *even* edges e_{2k} on *even* steps and *odd* edges e_{2k+1} on *odd* steps.

Let K_n denote the complete graph on n vertices. In this case, because K_n is so highly connected, the routing number of K_n is as small as one could hope for (see [7]):

$$rt(K_n) = 2 \ .$$

For the complete bipartite graph $K_{n,n}$ with $n \geq 3$, the following result is due to Wayne Goddard[143].

$$rt(K_{n,n}) = 4 \ .$$

For any tree T_n on n vertices, it was proved in [7] that

$$rt(T_n) < 3n.$$

However, the correct value of the constant may be half as large, as suggested by the following:

Conjecture. For any tree T_n on n vertices,

$$rt(T_n) \leq \lfloor \frac{3(n-1)}{2} \rfloor.$$

Furthermore, equality holds only when T_n is a star S_n on n vertices.

We remark that recently Louxin Zhang [255] proved an asymptotical version of the above conjecture by showing $rt(T_n) = 3n/2 + O(\log n)$.

The following result on routing on the hypercube can be traced back to the early work of switching networks (see Beneš [20]) and has appeared frequently in the literature on parallel computing:

$$rt(Q_n) \leq 2n - 1 \ .$$

The exact value of $rt(Q_n)$ is still unknown. It is easy to see that $rt(Q_n) \geq n$ since the diameter of Q_n is n. For small cases, it can be checked that $rt(Q_n) \geq n+1$ for $n = 2, 3$.

Problem: Is it true that for the n-cube Q_n,

$$rt(Q_n) = n + O(n) \, ?$$

Perhaps, $rt(Q_n) = n + o(n)$ for all sufficiently large n.

For the m by n grid graph $P_m \times P_n$, $m \leq n$,

$$rt(P_m \times P_n) \leq 2m + n .$$

In general, for the cartesian product of two graphs, we have [**7, 181**]

$$rt(G \times G') \leq 2rt(G) + rt(G') .$$

Note that since $G \times G'$ and $G' \times G$ are isomorphic graphs, this can be written in the symmetric form

$$rt(G \times G') \leq \min\{2rt(G) + rt(G'),\ 2rt(G') + rt(G).\}$$

Problem: Is it true that for every graph G,

$$rt(G \times G) \geq rt(G)?$$

From the above results and partial results, we can see that the problem of determining the routing number is quite difficult even for very special graphs. It is indeed surprising in a way that by using eigenvalues we can get very good approximations for the routing number problem. The following arguments are basically adapted from [**7**]. In the remaining part of this section, we assume G is a regular graph.

THEOREM 4.9. *Let G be a regular graph on n vertices and suppose $l \geq \log n/\lambda_1$. For each $v \in V$, independently, let $W(v)$ denote a random walk of length l starting at v. Let $I(v)$ denote the total number of other walks $W(u)$ such that there exists a vertex x and two indices $0 \leq i, j \leq l$, $|i - j| < 5$, so that $W(v)$ visits x at time i and $W(u)$ visits x at time j. Then, almost surely (i.e., with probability tending to 1 as n tends to infinity), there is no vertex v so that $I(v) > 100l$.*

The proof is very similar to that of Theorem 4.6 and is omitted.

THEOREM 4.10. *Let G denote a regular graph on n vertices and let σ be a permutation of order two on V (i.e., a product of pairwise disjoint transpositions). Put $l = \frac{10}{\lambda_1} \log n$. Then there is a set of n walks $W(v)$, $v \in V$, each of length $2l$, where both $W(v)$ and $W(\sigma(v))$ connect v and $\sigma(v)$ and traverse the same set of edges (in different directions) satisfying the following: If $I(v)$ denotes the total number of other walks $W(u)$ such that there exists a vertex x and two indices $0 \leq i, j \leq l$, $|i - j| < 5$, so that $W(v)$ visits x at time i and $W(u)$ visits x at time j or at time $2l - j$, then $I(v) \leq 400l$ for all v.*

THEOREM 4.11. *Let σ denote a permutation on the vertex set of G. Then*

$$rt(G, \sigma) = O(\frac{1}{\lambda_1} \log^2 n).$$

PROOF. Let G denote a regular graph on n vertices. It suffices to consider a permutation σ of order two on V since any permutation is a product of at most two such permutations (as proved in the proof of $rt(K_n) = 2$; also see [**7**]). We set $l = \frac{10}{\lambda_1} \log n$. We want to show that $rt(G, \sigma) = O(l^2)$. Let $W(v)$ be a system of walks of length $2l$ satisfying the assumptions of the previous theorem. Let H be the graph whose vertices are the walks $W(v)$ in which $W(u)$ and $W(v)$ are adjacent if there exists a vertex x and two indices $0 \leq i, j \leq l$, $|i - j| < 5$ so that $W(v)$ visits x at time i and $W(u)$ visits x at time j or at time $2l - j$. Then the maximum degree of H is $O(l)$ and hence it is $O(l)$-colorable. It follows that one can split all our paths $W(v)$ into $O(l)$ classes of paths such that the paths in each class are not adjacent in H. Consider now the following routing algorithm. For each set of paths as above, perform $2l$ steps, where the steps numbered i and $2l + 1 - i$ correspond to flipping the pebbles along edges numbered i and $2l + 1 - i$ in each of the paths in the set for all $1 \leq i \leq l$. One can check that by the end of these $2l$ steps, the ends of each path exchange pebbles, and all the other pebbles stay in their original places. (Note that some pebbles that are not at the ends of any of the paths may move several times during these steps, but the symmetric way these are performed guarantees that such pebbles will return to their original places at the completion of the $2l$ steps.) By repeating the above procedure for all the path-classes, the result follows.   □

We mention here several problems closely related to the routing number of a graph. One such problem is the following:

Suppose $G = (V, E)$ is a connected graph on n vertices. For a permutation π, we consider a *route set* P, which is just some set of paths P_i joining each vertex v_i to its destination vertex $\pi(v_i)$, for $i = 1, \cdots, n$. For each edge e of G, we consider the number $rc(e, G, \pi, P)$ of paths P_i in P which contain e. The *route covering number* $rc(G)$ of G is defined to be

$$rc(G) = \max_{\pi} \min_{P} \max_{e \in E} rc(e, G, \pi, P).$$

In other words, for each permutation we want to choose the route set so that the maximum number of occurrences of any edge in the paths of the route set is minimized.

For example, for the n-cube Q_n, the method [7] used to establish the route set gives

$$rc(Q_n) \leq 4.$$

In the other direction, by choosing π to be the permutation of vertices in Q_n so that the distance between v and $\pi(v)$ is n for every vertex v, it can be easily seen that

$$rc(Q_n) \geq \frac{\sum_v d(v, \pi(v))}{|E(Q_n)|} = 2.$$

Recently, Gu and Tamaki [**153**] proved that

$$rc(Q_n) = 2.$$

Also of interest is a "symmetric" version of the route covering problem:

A *pairing* for G is a partition of the vertex set of G into subsets of size 2. Is it possible to find edge-disjoint paths joining vertices of each pair for any pairing of G ?

4.5. Comparison theorems

We can often bound the eigenvalues of one graph by the eigenvalues of another provided pairs of adjacent vertices in the first graph can be joined by "short" paths in the second graph. Although the proofs for these comparison theorems are quite easy, the applications are abundant. Interesting examples along this line are given in Diaconis and Stroock [98] and numerous other papers [99, 100, 125] for comparing various different card shuffling schemes. We remark that the comparison theorems in this section can be viewed as generalized versions of the so-called "Poincaré" inequalities [98].

THEOREM 4.12. *Let G and G' be two connected regular graphs, with eigenvalues λ_1 and λ_1' and degrees k and k', respectively. Suppose that the vertex set of G is the same as the vertex set of G'. We assume that for each edge $\{x, y\}$ in G, there is a path $P(x, y)$ in G' joining x and y of length at most l. Furthermore, suppose that every edge in G' is contained in at most m paths $P(x, y)$. Then we have*

$$\lambda_1' \geq \frac{k\lambda_1}{k'lm}.$$

PROOF. Using the definition of the eigenvalues, we consider the harmonic eigenfunction f achieving λ_1 in G'.

$$
\begin{aligned}
\lambda_1' &= \frac{\displaystyle\sum_{\{x,y\}\in E(G')} (f(x)-f(y))^2}{k'\displaystyle\sum f^2(x)} \\
&= \frac{k\displaystyle\sum_{\{x,y\}\in E(G')}(f(x)-f(y))^2}{k'\displaystyle\sum_{\{x,y\}\in E(G)}(f(x)-f(y))^2} \cdot \frac{\displaystyle\sum_{\{x,y\}\in E(G)}(f(x)-f(y))^2}{k\displaystyle\sum f^2(x)}
\end{aligned}
$$

We note that for $\{x, y\} \in E(G)$ and path $P(x, y)$ joining x and y in G', we have

$$(f(x)-f(y))^2 \leq |P(x,y)| \sum_{e\in P(x,y)} f^2(e) \leq l\sum_{e\in P(x,y)} f^2(e)$$

where $f^2(e) = (f(x)-f(y))^2$ for $e = \{x, y\}$, and $|P(x, y)|$ denotes the number of edges of G' in $P(x, y)$. Hence

$$
\begin{aligned}
m\sum_{e\in E(G')} f^2(e) &\geq \sum_{\{x,y\}\in E(G)}\sum_{e\in P(x,y)} f^2(e) \\
&\geq \frac{1}{l}\sum_{\{x,y\}\in E(G)} (f(x)-f(y))^2.
\end{aligned}
$$

Therefore we have

$$\lambda_1' \geq \frac{k}{k'lm} \cdot \frac{\sum\limits_{\{x,y\} \in E(G)} (f(x) - f(y))^2}{\sum f^2(x)k}$$

$$\geq \frac{k}{k'lm} \lambda_1.$$

This completes the proof of Theorem 4.12.   □

It is not surprising that the above proof is quite similar to some of those in Section 4.2. There are several generalizations of Theorem 4.12:

THEOREM 4.13. *Let G and G' be two connected graphs, with eigenvalues λ_1 and λ_1', respectively. Suppose that the vertex set of G is the same as the vertex set of G'. Assume that for each edge $\{x, y\}$ in G, there is a path $P(x, y)$ in G' of length at most l, and for each vertex v, the degree d_v of v in G is at least ad_v', where d_v' is the degree of v in G'. Furthermore, suppose every edge in G' is contained in at most m paths $P(x, y)$. Then we have*

$$\lambda_1' \geq \frac{a\lambda_1}{lm}.$$

Instead of proving Theorem 4.13, we will prove the following generalization:

THEOREM 4.14. *Let G and G' be two connected graphs, with eigenvalues λ_1 and λ_1', respectively. Suppose that the vertex set of G can be embedded into the vertex set of G' under the mapping $\varphi : V(G) \to V(G')$. Suppose φ satisfies the following conditions for fixed positive values a, l, m:*

(a): *Each edge $\{x, y\}$ in $E(G)$ is associated with a path, denoted by $P_{x,y}$, joining $\varphi(x)$ to $\varphi(y)$ in G' of length at most l.*
(b): *Let d_v, d_v' denote the degrees of v in G and in G', respectively. For any v in $V(G')$, we have*

$$\sum_{x \in \varphi^{-1}(v)} d_x \geq ad_v'.$$

(c): *Each edge in G' is contained in at most m paths $P_{x,y}$.*

Then we have

$$\lambda_1' \geq \frac{a\lambda_1}{lm}.$$

PROOF. The proof is very similar to that of Theorem 4.12. For a harmonic eigenfunction g of G', we define $f : V(G) \to \mathbb{R}$ as follows: For a vertex x in $V(G)$,

$$f(x) = g(\varphi(x)) - c$$

where the constant c is chosen to satisfy

$$\sum_x f(x)d_x = 0.$$

We note that

$$
\begin{aligned}
\sum_{x \in V(G)} f^2(x) d_x &= \sum_{x \in V(G)} (g(\varphi(x)) - c))^2 d_x \\
&= \sum_{v \in V(G')} (g(v) - c)^2 \sum_{\varphi^{-1}(v) = x} d_x \\
(4.1) \qquad\qquad &\geq a \sum_{v \in V(G')} g(v)^2 d_v'.
\end{aligned}
$$

Now, for $\{x, y\} \in E(G)$ with $\varphi(x) = u, \varphi(y) = v$, let $P_{x,y}$ denote the path corresponding to $\{x, y\}$ joining u and v in G'. We have

$$
(g(u) - g(v))^2 \leq |P_{x,y}| \sum_{e \in P_{x,y}} g^2(e) \leq l \sum_{e \in P_{x,y}} g^2(e),
$$

where $g^2(e) = (g(a) - g(b))^2$ for $e = \{a, b\}$. Hence we have

$$
\begin{aligned}
m \sum_{e \in E(G')} g^2(e) &\geq \sum_{e \in E(G')} m g^2(e) \\
&\geq \sum_{\{x,y\} \in E(G)} \sum_{e \in P(x,y)} g^2(e) \\
&\geq \sum_{\substack{\{x,y\} \in E(G) \\ u = \varphi(x) \\ v = \varphi(y)}} \frac{1}{l} (g(u) - g(v))^2 \\
(4.2) \qquad\qquad &\geq \frac{1}{l} \sum_{\{x,y\} \in E(G)} (f(x) - f(y))^2.
\end{aligned}
$$

Combining inequalities $(4.1),(4.2)$, we have

$$
\begin{aligned}
\lambda_1' &= \sup_t \frac{\displaystyle\sum_{\{u,v\} \in E(G')} (g(u) - g(v))^2}{\displaystyle\sum_{v \in V(G')} (g(v) - t)^2 d_v'} \\[2mm]
&\geq \frac{\displaystyle\sum_{\{u,v\} \in E(G')} (g(u) - g(v))^2}{\displaystyle\sum_{v \in V(G')} (g(v) - c)^2 d_v'} \\[2mm]
&= \frac{\displaystyle\sum_{\{u,v\} \in E(G')} (g(u) - g(v))^2}{\displaystyle\sum_{\{x,y\} \in E(G)} (f(x) - f(y))^2} \cdot \frac{\displaystyle\sum_{\{x,y\} \in E(G)} (f(x) - f(y))^2}{\displaystyle\sum_{x \in V(G)} f^2(x) d_x} \cdot \frac{\displaystyle\sum_{x \in V(G)} f^2(x) d_x}{\displaystyle\sum_{v \in V(G')} (g(v) - c)^2 d_v'} \\[2mm]
&\geq \frac{1}{ml} \cdot \frac{\displaystyle\sum_{\{x,y\} \in E(G)} (f(x) - f(y))^2}{\displaystyle\sum_{x \in V(G)} f^2(x) d_x} a
\end{aligned}
$$

by using

$$\sum_{x \in \varphi^{-1}(v)} f^2(x) d_x \geq (g(v) - c)^2 d'_v.$$

Since

$$\sum_x f(x) d_x = 0,$$

we have

$$\frac{\displaystyle\sum_{\{x,y\} \in E(G)} (f(x) - f(y))^2}{\displaystyle\sum_{x \in V(G)} f^2(x) d_x} \geq \lambda_1.$$

Hence

$$\lambda'_1 \geq \frac{a}{ml} \lambda_1$$

and the proof of Theorem 4.14 is complete. □

We remark that Theorems 4.3 and 4.5 in Section 4.2 are just special cases of Theorem 4.14 in which G is taken to be a complete graph and G' is chosen arbitrarily.

We also remark that the generalized version in Theorem 4.14 can often give stronger results for certain problems. For example, we consider the following simple and natural random walk problem on generating sets of groups which arises in computational group theory.

EXAMPLE 4.15. Let H denote a graph on n vertices each of which is labelled by an element of a group Γ. At each unit of time, one of the vertices, say v with label g, can be changed to gf where f or f^{-1} is a label of a neighbor u of v. Suppose we start with the case that the set of all vertex labels generates the group Γ. The problem of interest is to determine how rapidly this processes mixes, i.e., how many steps it requires to be close to a "random" generating set.

By using Theorem 4.14, we can obtain an upper bound of the form cDn^2, where c depends only on the size of Γ, and D denotes the diameter of H (see [80]). Similar bounds have also been obtained by Diaconis and Saloff-Coste in [102] using more complex comparison techniques. However, all of these bounds are rather far from what is believed to be the truth, namely, that order $n \log n$ steps (under total variation distance). In [81], it is proved that in fact this bound is achieved for the case that $H = \mathbb{Z}_2$. Interestingly, for relatively pointwise distance, $O(n^2)$ is proved to be the correct bound [81].

Notes

Path arguments has been used early on to compute isoperimetric constants of a graph. For example, in the early work of Bhatt and Leighton [33] on VLSI design and parallel computation, path arguments were extensively utilized. Jerrum and Sinclair [**167**] used path arguments to bound the Cheeger constant in order to bound the eigenvalues in their seminal work of estimating permanents. Diaconis and Stroock [**98**] used path arguments to directly bound eigenvalues.

Sections 4.3 and 4.4 on paths and routing are mainly based on [**7**]. Some variations of the comparison theorems in Sections 4.2 and 4.5 can be found in [**80**].

CHAPTER 5

Eigenvalues and quasi-randomness

5.1. Quasi-randomness

The isoperimetric properties and eigenvalue inequalities in previous sections are quite effective in dealing with many aspects of graphs with λ_1 bounded away from 0, i.e., $\lambda_1 \geq c > 0$ for some absolute constant c. Basically, such graphs have "small" diameters and "large" vertex and edge expansions. Furthermore, there are "many" paths with "small" congestion joining given pairs of vertices as well as various dynamic routing schemes for sending messages (or pebbles) along the paths simultaneously. The main questions concern how "small" or "large" these quantities can be. Many answers lie in estimating these quantities in terms of eigenvalues.

When all λ_i are close to 1, the graphs satisfy additional nice properties which will be discussed in this chapter. For example, a random graph has eigenvalues $|1 - \lambda_i| = O(\frac{1}{\sqrt{n}})$ for $i \neq 0$. (Here we can use any model of a random graph [30]). In other words, almost all graphs satisfy $|1 - \lambda_i| = O(\frac{1}{\sqrt{n}})$ for all $i \neq 0$. For sparse graphs with average degree k, the corresponding condition is $|1 - \lambda_i| = O(\frac{1}{\sqrt{k}})$.

There is a long history in extremal graph theory for investigating how one graph property relates to another. In recent years, there has been a great deal of development in the study of random graphs, which analyzes the behavior of almost all graphs. These two areas which might seem to be quite different, have a great deal of overlap in the following sense: Most graph properties which are satisfied by a random graph are related in a strong way. There is a series of papers [67] [70] [71] [72] [73] [74] [75] [76] [227] introducing a large *equivalence class* of graph properties, in the sense that any graph which satisfies any one of the properties must satisfy all of them. Graphs that satisfy these properties are said to be *quasi-random*. One of the main quasi-random properties concerns the concentration of all nontrivial eigenvalues. So, such an eigenvalue distribution dictates much of the behavior of the graph.

As an example, here we illustrate the following list of quasi-random properties for a graph G with edge density $1/2$ [76]:

P_1 : $\max\limits_{i \neq 0} |1 - \lambda_i| = o(1)$ and G is almost regular (i.e., all except $o(n)$ vertices

have degree $n/2 + o(n)$).

P_2 :   For each subset $S \subseteq V(G)$, the number $e(S)$ of edges of G with both endpoint in S satisfies

$$e(S) = \tfrac{1}{4}|S|^2 + o(n^2).$$

P_3 :   For each subset $S \subseteq V(G)$ with $|S| = \lfloor \tfrac{n}{2} \rfloor$,

$$e(S) = (\tfrac{1}{16} + o(1))n^2.$$

P_4 :   For u in $V(G)$, let $N(u)$ denote the neighborhood $\{v : u \sim v\}$ of u.

$$\sum_{u,v} |N(u) \cap N(v) - \frac{n}{4}| = o(n^2).$$

$P_5(s)$:   For $s \geq 4$ and for all graphs $M(s)$ on s vertices, the number $N_G^*(M(s))$ of "labelled" induced subgraphs of G isomorphic to $M(s)$ satisfies

$$N_G^*(M(s)) = (1 + o(1))n^s 2^{-\binom{s}{2}}.$$

$P_6(t)$:   For the $2t$-cycle C_{2t}, $t \geq 2$, the number $N_G(C_{2t})$ of occurrences of C_{2t} as a (labelled) subgraph of G is

$$N_G(C_{2t}) = (1 + o(1))(\tfrac{n}{2})^t,$$

and $e(G) \geq (1 + o(1))\frac{n^2}{4}.$

We remark that the description of our graph properties typically contain the $o(\cdot)$ notation. The statement $P(o(1)) \Rightarrow P'(o(1))$ means that for any $\epsilon > 0$, there exists δ such that $P(\delta) \Rightarrow P'(\epsilon)$. Two properties P and P' are equivalent if $P \Rightarrow P'$ and $P' \Rightarrow P$. In [76], it was shown that all P_i, $i = 1 \cdots 6$, are equivalent.

The list of quasi-random properties is still increasing (see [227]) and each new addition further strengthens the strong consequences of the equivalence. Although some properties are easy to compute (such as P_1, P_4, P_6), some are (at present) computationally intractable (e.g. P_2, P_3, P_5). We can construct graphs satisfying all the properties by verifying only one property. Thus, this provides a validation scheme for approximating one difficult property by using another equivalent property which is easier to compute. It is also desirable to have a quantitative estimate for the error in the $o(\cdot)$ term, which is often the focus of many extremal graph problems.

The main goal of this chapter is to give a general unified treatment of quasi-random graphs. Throughout, the eigenvalue property will be central and the goal will be to bound the $o(1)$ estimates in each of the properties in terms of $\bar{\lambda} = \max_{i \neq 0} |1 - \lambda_i|$, if possible. We will examine several major properties with emphasis on their relations to the eigenvalues.

5.2. The discrepancy property

Let G denote a graph having a vertex set V with n vertices and an edge set E with e edges. The edge density ρ is defined to be $2e/n^2$. For subsets $X, Y \subset V$, we recall the notation $E(X, Y)$ as the set of ordered pairs corresponding to edges with one endpoint in X and the other in Y, i.e.,

$$E(X, Y) = \{(u, v) : u \in X, v \in Y \text{ and } \{u, v\} \in E\}.$$

Here, X and Y are not necessarily disjoint. We denote

$$e(X, Y) = |E(X, Y)|.$$

For a subset S of V, the discrepancy of S, denoted by $disc(G, S)$, is defined to be

$$disc(G, S) = |e(S, S) - \rho|S|^2|$$

The α-discrepancy of G is the maximum discrepancy of $S \subseteq V$ over all S with $|S| = \alpha n$, where $0 < \alpha \le 1$, i. e.,

$$disc(G; \alpha) = \max_{|S| = \lfloor \alpha n \rfloor} disc(G, S).$$

In particular, the discrepancy of a graph G, denoted by $disc\ G$ is just

$$disc\ G = \max_{\alpha} disc(G; \alpha).$$

In a certain sense, the discrepancy is the "quantitative" version of the Ramsey property which asserts that when α is very small $\left(\sim \frac{c \log n}{n} \right)$, the α-discrepancy can be as large as $\rho|S|^2$. In general, the problem of determining the α-discrepancy is a very difficult problem and is known to be NP-complete. It is therefore of interest to derive upper bounds for the discrepancy using other methods, e.g., by eigenvalue arguments.

For a subset S of the vertex set V of G, we consider the characteristic vector ψ_S, defined by

$$\psi_S(u) = \begin{cases} 1 & \text{if } u \in S, \\ 0 & \text{otherwise.} \end{cases}$$

We note that

$$\langle \psi_S, A\psi_S \rangle = \sum_{u \in S} \sum_{v \in S} A_{uv} = e(S, S)$$

where A is the adjacency matrix of G.

Also, the edge density satisfies

$$\rho = \frac{\langle \mathbf{1}, A\mathbf{1} \rangle}{\langle \mathbf{1}, J\mathbf{1} \rangle}$$

and

$$|S| = \langle \psi_S, \mathbf{1} \rangle$$

where J denotes the all 1's matrix.

If $\mathbf{1}$ is an eigenvector of A (as is the case for regular graphs), we can then successfully bound the discrepancy by the bounds on the eigenvalues of the adjacency matrix. However, for general graphs, $\mathbf{1}$ is usually not an eigenvector for A. These obstacles can be overcome by considering the eigenvalues of the Laplacian as in the following:

THEOREM 5.1. *Suppose X, Y are two subsets of the vertex set V of a graph G. Then*

$$|e(X, Y) - \frac{\text{vol } X \text{ vol } Y}{\text{vol } G}| \leq \bar{\lambda}\sqrt{\text{vol } X \text{ vol } Y}$$

where $\bar{\lambda} = \max_{i \neq 0} |1 - \lambda_i|$.

PROOF. Following the above notation, we have

$$e(X, Y) = \psi_X \, T^{1/2}(I - \mathcal{L})T^{1/2} \, \psi_Y^*.$$

Suppose

$$T^{1/2} \, \psi_X = \sum_i a_i \phi_i,$$

$$T^{1/2} \, \psi_Y = \sum_i b_i \phi_i,$$

where ϕ_i's are eigenvectors of $\mathcal{L}$ and, in particular, $\phi_0 = T^{1/2} \, \mathbf{1}/\sqrt{\text{vol } G}$.

Since ϕ_i, $i \geq 1$, is orthogonal to ϕ_0, we have

$$JT^{1/2}\phi_i = 0$$

for all $i \geq 1$.

Also,

$$\text{vol } X \text{ vol } Y = \psi_X \, T^{1/2} \, (T^{1/2}JT^{1/2}) \, T^{1/2} \, \psi_Y^*.$$

Therefore, we have

$$
\begin{aligned}
disc(X, Y) &= |e(X, Y) - a_0 b_0| \\
&= |\psi_X \, A \, \psi_Y^* - a_0 b_0| \\
&= |\psi_X \, T^{1/2} \, (I - \mathcal{L} - \frac{T^{1/2}JT^{1/2}}{\text{vol } G}) \, T^{1/2} \, \psi_Y^*| \\
&= |\sum_{i \geq 1} a_i b_i (1 - \lambda_i)| \\
&\leq \bar{\lambda}\sqrt{\sum_{i \geq 1} a_i^2 \sum_{i \geq 1} b_i^2} \\
&\leq \bar{\lambda}\frac{\sqrt{\text{vol } X \text{ vol } \bar{X} \text{ vol } Y \text{ vol } \bar{Y}}}{\text{vol } G}
\end{aligned}
$$

since

$$\sum_i a_i^2 = \text{vol } X$$

$$a_0 = \langle \phi_0, T^{1/2} \, \psi_X \rangle = \frac{\text{vol } X}{\sqrt{\text{vol } G}}$$

$$\sum_{i \geq 1} a_i^2 = \text{vol } G - \frac{(\text{vol } X)^2}{\text{vol } G} = \frac{\text{vol } X \text{ vol } \bar{X}}{\text{vol } G} \leq \text{vol } X.$$

This completes the proof of this theorem.   □

In fact, we have shown a slightly stronger result:

THEOREM 5.2. *Suppose X, Y are two subsets of the vertex set V of a graph G. Then*

$$(5.1) \qquad |e(X,Y) - \frac{\text{vol } X \text{ vol } Y}{\text{vol } G}| \leq \bar{\lambda} \frac{\sqrt{\text{vol } X \text{ vol } Y \text{vol } \bar{X} \text{ vol } \bar{Y}}}{\text{vol } G}$$

where $\bar{\lambda} = \max_{i \neq 0} |1 - \lambda_i|$.

Here we state a few consequences of Theorem 5.1:

COROLLARY 5.3. *Suppose X is a subset of vertices in a graph G. Then*

$$|e(X,X) - \frac{(\text{vol } X)^2}{\text{vol } G}| \leq \bar{\lambda} \frac{\text{vol } X \text{ vol } \bar{X}}{\text{vol } G} \leq \bar{\lambda} \text{vol } X.$$

COROLLARY 5.4. *Suppose X is a subset of vertices in a k-regular graph G. Then*

$$|e(X,X) - \frac{k|X|^2}{n}| \leq k\bar{\lambda}|X|.$$

We note that Theorem 5.2 is closely related to the edge-expansion properties that we discussed in Chapter 2. If we can find an upper bound for the number of edges both ends of which lie inside a set X, then we can find a lower bound for the number of edges leaving X. Using Theorem 5.2 and Corollary 5.3, we have the following isoperimetric inequality.

COROLLARY 5.5. *Suppose X is a subset of vertices in a graph G. Then the edge boundary ∂X satisfies*

$$\frac{|\partial X|}{\text{vol } X} \geq (1 - \bar{\lambda}) \frac{\text{vol } \bar{X}}{\text{vol } G}.$$

PROOF. We consider

$$\text{vol } X = e(X,X) + |\partial X|.$$

By substituting using Corollary 5.3, we have

$$\frac{|\partial X|}{\text{vol } X} \geq (1 - \bar{\lambda}) \frac{\text{vol } \bar{X}}{\text{vol } G}.$$

□

From the statement of (5.1), it is tempting to make a number of conjectures. Some of these questions can be partially answered, but most of them are unresolved.

Question 1: Suppose a graph G satisfies, for some fixed α,

$$|e(X,Y) - \frac{\text{vol } X \text{ vol } Y}{\text{vol } G}| \leq \alpha \frac{\sqrt{\text{vol } X \text{ vol } Y \text{ vol } \bar{X} \text{ vol } \bar{Y}}}{\text{vol } G}$$

for all $X, Y \subseteq V(G)$.

Is it then true that $\bar{\lambda} \leq 100\alpha$?
(Of course, 100 can be replaced by 10^{100}, if this is easier.)

Suppose we use Corollary 5.5 so that we get $h_G \geq \frac{1-\alpha}{2}$. Using Cheeger's inequality from Chapter 2, we have

$$\lambda_1 \geq \frac{h_G^2}{2} \geq \frac{(1-\alpha)^2}{8}.$$

We therefore have shown that

$$1 - \lambda_1 \leq 1 - \frac{(1-\alpha)^2}{8}.$$

This is still quite distant from the desired bound $\bar{\lambda} \leq 100\alpha$. However, it is evidence in support of an affirmative answer to the above question.

Question 2: Suppose a graph G satisfies, for some fixed α,

$$(5.2) \qquad |e(X,Y) - \frac{\text{vol } X \text{ vol } Y}{\text{vol } G}| \leq \alpha \frac{\sqrt{\text{vol } X \text{ vol } Y \text{ vol } \bar{X} \text{ vol } \bar{Y}}}{\text{vol } G}$$

for all $X, Y \subseteq V(G)$.

Is it true that disc $G \leq 100\alpha$?

In other words, for all $X, Y \subseteq V(G)$, does the following equality hold:

$$|e(X,Y) - \rho|X|\,|Y|\,| \leq 100\alpha \frac{\sqrt{\text{vol } X \text{ vol } Y \text{ vol } \bar{X} \text{ vol } \bar{Y}}}{\text{vol } G}$$

where $\rho = \frac{\text{vol } G}{n^2}$ is the edge density of G?

This question can be answered in the negative. We now construct a graph satisfying the above assumption but having a large discrepancy. In fact, the graph is not even "almost regular."

Let H be a graph with vertex set $A \cup B$ where $|A| = \frac{n}{2}$, $|B| = \frac{n}{2}$ and $A \cap B = \emptyset$. The induced subgraph on A will be a random graph with edge density $1/2$. The induced subgraph on B will be a random graph with edge density $1/4$. The bipartite subgraph between A and B will have edge density $1/3$. It is left as an exercise to check that H satisfies (5.2), but the discrepancy is large (about cn^2 where $c \geq 0.01$).

It is of course true that if a graph is "almost regular" and it satisfies (5.2), then its discrepancy is small. "Almost regular" is a necessary condition for quasi-randomness. In the fourth section of this chapter, we will examine how the discrepancy relates to other quasi-random properties.

Theorem 5.2 gives a good approximation for the discrepancy of many families of graphs. For example, many expander graphs (see Chapter 6) and the random graphs (as described in Remark 1) have eigenvalues

$$\bar{\lambda} \sim \frac{1}{\sqrt{k}}$$

where k is the average degree. For such graphs G, the above theorem implies

$$disc(G; \alpha) := \sup_{|X| = \alpha n} |e(X, X) - \rho |X|^2| \leq c \sqrt{k} \alpha n$$

for some absolute constant c.

In the book of Erdős-Spencer [120], a lower bound for the discrepancy of any graph G with edge density $1/2$ was given by

$$disc\ G \geq cn^{3/2}$$

for some absolute constant c. So, the eigenvalue upper bound for the discrepancy is within a constant factor of the best possible value for many graphs.

The discrepancy of a random graph can be easily estimated. Here is a sketch of a proof that

$$disc(G; \alpha) \sim cn^{3/2}$$

for a random graph G on n vertices where c is a constant depending only on the edge density. Here we assume that the edge density ρ is a fixed positive quantity when n approaches infinity.

Let G denote a random graph with edge density ρ. We define a function f which assigns the value $(1 - \rho)$ to every edge of G and the value $-\rho$ to every non-edge of G. It is easy to see that $| \sum_{u,v \in S} f(u, v) |= disc(G, S)$. Using the Chernoff bound (see [12], pp. 237), the probability that the random graph has discrepancy more than β satisfies

$$Prob(disc(G; \alpha) > \beta) \leq exp(-\beta^2 / (2\rho \alpha^2 n^2)).$$

Therefore the total probability of having some set of size αn for which the discrepancy is β is at most

$$\binom{n}{\alpha n} exp(-\beta^2 / (2\rho \alpha^2 n^2)).$$

When the above quantity is smaller than 1, there must exist a graph with discrepancy no more than β. Indeed, we can choose β to be $c\,\alpha n^{3/2}$ for some appropriate constant c so this is true. The exact expression of c in terms of α and ρ are not hard to derive. (Hint, try $c = \alpha \sqrt{\rho H(\alpha)}$ where $H(x)$ is the "entropy" function.)

How will the discrepancy $disc(G; \alpha)$ behave when α is small, say, for example, αn is smaller than $\sqrt{n}$ for a graph with edge density $1/2$? This, in fact, embodies a wide collection of classical combinatorial problems for the whole range of subset sizes αn, say, from 0 to n, and/or for any edge density. For most of such problems, our knowledge is quite limited and the known tools are few. Perhaps the only powerful method is to use eigenvalues to upper bound the discrepancy for regular graphs when $\alpha > \bar{\lambda}$ as demonstrated in Corollary 5.4. For general α, such as $\alpha < \bar{\lambda}$, the discrepancy

$$disc(n; \alpha) = \sup_{|V(G)| = n} disc(G; \alpha)$$

is not well understood.

It would be tempting to define discrepancy as

$$\sup_X \frac{|\, e(X,X) - \rho|X|^2\,|}{|X|^2}$$

However, from classical Ramsey theory we know that any graph on n vertices contains an induced subgraph on a subset X of $c \log n$ vertices which is either a complete subgraph or an independent set. Thus, $|e(X,X) - \rho|X|^2|$ could be as large as $\rho|X|^2$. In a way, discrepancy problems can be viewed as a qualitative generalization of Ramsey theory. The discrepancy is concerned with induced subgraphs of all sizes while Ramsey theory focuses on the containment of special subgraphs (which can be small subgraphs with large discrepancies).

Most bounds like the ones above are proved by using probabilistic methods. Explicit constructions are quite poor in their performance in comparison with the random graphs. The reader is referred to several excellent papers and surveys by Thomason [237, 238, 239] on this subject under the name of "$(p, \alpha) - jumbled$" graphs. For the lower range of α, the discrepancy problems are basic Ramsey problems which we will discuss briefly in the next subsection.

5.2.1. The Ramsey property. A fundamental result of Ramsey [218] guarantees the existence of a number $R(k, \ell)$ so that any graph on $n \geq R(k, \ell)$ vertices contains either a complete graph of size k or an independent set of size ℓ. The problem of determining $R(k, \ell)$ is notoriously difficult. The first non-trivial lower bound for $R(k, k)$, due to Erdős [111] in 1947, states

$$(5.3) \qquad R(k,k) > (1 + o(1)) \frac{1}{e\sqrt{2}}\, k \cdot 2^{k/2}.$$

In other words, there exist graphs on n vertices which contain no cliques or independent sets of size $2 \log n$ when n is sufficiently large. The proof for (5.3) is simple and elegant, and is based on the observation that the probability of having a clique or independent set of size k is at most $\binom{n}{k} \cdot 2^{1 - \binom{k}{2}}$. We see that if this quantity is less than one, there must exist a graph without any clique or independent set of size k.

This basic result plays an essential role in laying the foundations for both Ramsey theory and probabilistic methods, two of the major thriving areas in combinatorics. In the 40 years since its proof, the bound in (5.3) has only been improved by a factor of 2, again by probabilistic arguments [232].

Attempts have been made over the years to construct good graphs (i.e., with small cliques and independent sets) without much success [74, 144]. H.L Abbott [1] gave a recursive construction with cliques and independence sets of size $cn^{\log 2/ \log 5}$. Nagy [204] gave a construction reducing the size to $cn^{1/3}$. A breakthrough finally occurred several years ago with the result of Frankl [128], who gave the first Ramsey construction with cliques and independent sets of size smaller than $n^{1/k}$ for any k. This was further improved to $e^{c(\log n)^{3/4}(\log \log n)^{1/4}}$ in [64]. Here we will outline a construction of Frankl and Wilson [130] for Ramsey graphs with cliques and independent sets of size at most $e^{c(\log n \log \log n)^{1/2}}$.

EXAMPLE 5.6. Let q be a prime power. The graph G will have vertex set $V = \{F \subseteq \{1, \cdots, m\} : \mid F \mid = q^2 - 1\}$ and edge set $E = \{(F, F') : \mid F \cap F' \mid \not\equiv -1 \,(\mathrm{mod}\ q)\}$. A result in [130] implies that G contains no clique or independent set of size $\binom{m}{q-1}$. By choosing $m = q^3$, we obtain a graph on $n = \binom{m}{q^2 - 1}$ vertices containing no clique or independent set of size $e^{c(\log n \, \log \log n)^{1/2}}$.

The proof involved in the above construction is based on a beautiful result of Ray-Chaudhuri and Wilson [220] on intersection theorems. This type of intersection graph and the related intersection theorems provide excellent examples for many extremal problems including the discrepancy problems.

A graph which has often been suggested as a natural candidate for a Ramsey graph is the Paley graph (see more discussion in Section 6). Very little is known about its maximum size of cliques and independent sets. For the lower bound, a result of S. Graham and C. Ringrose [145] shows that infinitely many Paley graphs on p vertices contain a clique of size $c \log p \log \log \log p$. (This contrasts with the trivial upper bound of $c\sqrt{p}$.) Earlier results of Montgomery [202] show that assuming the Generalized Riemann Hypothesis, we would have a lower bound $c \log p \log \log p$ infinitely often. If we take the Ramsey property as a measure of "randomness," the above results show that the Paley graphs deviate from random graphs. There is no question that the problem in constructive methods for which a solution is most widely sought is the following, posed long ago (as early as the 40's) by Erdős:

Problem : Construct graphs on n vertices containing no clique and no independent set of size $c \log n$.

Instead of focusing on the occurrence of cliques and independence sets, similar problems can be considered on the occurrence or the frequency of other specified subgraphs [35, 146, 222, 252]. It is not difficult to show that almost all graphs contain every graph with at most $2 \log n$ vertices as an induced subgraph. The best current constructions containing every graph with up to $c\sqrt{\log n}$ vertices as induced subgraphs can be found in [73, 129].

5.3. The deviation of a graph

We have discussed various aspects of the discrepancy of a graph. In spite of the important role discrepancy plays in various extremal problems, one major challenge is that discrepancy is difficult to compute since its definition involves taking the extremum over all choices of subsets with potentially exponentially many cases. Here we will consider another invariant, the so-called deviation of a graph. Although its definition seems to be more complicated, it can be easily computed in polynomial time. Furthermore, deviation is very closely related to discrepancy and it can be used to prove upper and lower bounds for discrepancy.

For a graph G with edge density ρ, we define a weighted indicator function $\chi : V \times V \to \mathbb{R}$ as follows:

$$\chi(x,y) = \begin{cases} 1 - \rho & \text{if } x \sim y \\ -\rho & \text{otherwise.} \end{cases}$$

For a 4-cycle C, we denote

$$\chi(C) = \prod_{\{x,y\} \in C} \chi(x,y)$$

and we define the deviation of G by

$$
\begin{aligned}
dev \ G &= \frac{1}{\rho^4 n^4} \sum_C \chi(C) \\
&= \frac{1}{\rho^4 n^4} \sum_{x,y,z,w} \chi(x,y)\chi(y,z)\chi(z,w)\chi(w,x)
\end{aligned}
$$

where x,y,z,w range independently over all vertices of G. For the special case of $\rho = 1/2$, the deviation is exactly $1/16n^4$ times the number of "even" 4-cycles minus the number of "odd" 4-cycles. (A 4-cycle $\{x,y,z,w\}$ is said to be "even" if $\chi(x,y)\chi(y,z)\chi(z,w)\chi(w,x)$ is positive.)

Before we derive relations between deviation and eigenvalues, we want to give a quantitative measure of regularity and irregularity in a graph. (Recall that "almost regular" means that all but $o(n)$ vertices have degree within $(1+o(1))$ of the average degree of the graph). We define the *irregularity* of a graph G, denoted by $irr \ G$ as follows:

$$irr \ G := \frac{\rho}{n} \sum_{v \in V(G)} (\frac{1}{\rho_v} - \frac{1}{\rho})$$

where $|V(G)| = n$, $\rho_v = d_v/n$, and $\rho = \sum d_v / n^2$.

The smaller the value of $irr \ G$ is, the more closely the graph G approximates a regular graph. When $irr \ G = 0$, G is regular, as shown by the following useful lemma. The discussions and techniques in this section can be greatly simplified if we only consider regular graphs. However, in the same spirit of preceding sections, we consider a general graph for completeness.

LEMMA 5.7. *For any $A \subseteq V(G)$, we have*

$$\sum_{v \in V(G)} (\frac{1}{\rho_v} - \frac{1}{\rho}) \geq \frac{(\text{vol } A - |A|\rho n)^2}{\rho^2 n \text{ vol } A} \quad \text{for any } A \subseteq V(G).$$

PROOF.

$$\sum_{v \in V(G)} (\frac{1}{\rho_v} - \frac{1}{\rho}) = \sum_{v \in V(G)} \frac{1}{\rho_v}(1 - \frac{\rho_v}{\rho})^2$$

$$\geq \sum_{v \in A} \frac{1}{\rho_v}(1 - \frac{\rho_v}{\rho})^2$$

$$= \sum_{v \in A}(\frac{1}{\rho_v} - \frac{2}{\rho} + \frac{\rho_v}{\rho^2})$$

$$= \frac{\text{vol } A - |A|\rho n}{\rho^2 n} + \sum_{v \in A}(\frac{1}{\rho_v} - \frac{1}{\rho}).$$

For $\rho_A = \sum_v \rho_v/|A|$, we have

$$\sum_{v \in A}(\frac{1}{\rho_v} - \frac{1}{\rho_A}) \geq 0.$$

Therefore,

$$\sum_{v \in A}(\frac{1}{\rho_v} - \frac{1}{\rho}) = \sum_{v \in A}(\frac{1}{\rho_v} - \frac{1}{\rho_A} + \frac{1}{\rho_A} - \frac{1}{\rho})$$

$$\geq \sum_{v \in A}(\frac{1}{\rho_A} - \frac{1}{\rho})$$

$$= |A| \cdot (\frac{1}{\rho_A} - \frac{1}{\rho})$$

$$= |A|(\frac{|A|n}{\text{vol } A} - \frac{1}{\rho})$$

$$= \frac{(|A|\rho n - \text{vol } A)}{\rho \text{ vol } A}|A|.$$

Therefore we have

$$\sum_{v \in V(G)}(\frac{1}{\rho_v} - \frac{1}{\rho}) \geq \frac{(\text{vol } A - |A|\rho n)^2}{\rho^2 n \text{ vol } A}.$$

The lemma is proved. $\square$

The above lemma implies the following useful facts:

COROLLARY 5.8.

$$| \text{ vol } A - |A|\rho n | \leq \sqrt{irr \ G} \text{ vol } G$$

for all $A \subseteq V(G)$.

LEMMA 5.9.

$$\sup_{A \subseteq V(G)} |\sum_{v \in A}(\frac{1}{\rho_v} - \frac{1}{\rho})| \leq \frac{n}{\rho}(irr \ G + \sqrt{irr \ G})$$

PROOF. From the proof of Lemma 5.7, we have

$$\sum_{v \in A}(\frac{1}{\rho_v} - \frac{1}{\rho}) \leq \sum_{v \in V(G)}(\frac{1}{\rho_v} - \frac{1}{\rho}) + \frac{|A|\rho n - \text{vol } A}{\rho^2 n}$$

$$\leq \frac{n}{\rho}irr \ G + \frac{\sqrt{\rho n^2 \text{vol } A \ irr \ G}}{\rho^2 n}$$

$$= \frac{n}{\rho}irr \ G + \frac{n}{\rho}\sqrt{irr \ G}.$$

Therefore, Lemma 5.9 is proved. $\qquad\square$

The deviation $dev \ G$ can be expressed in the following form.

THEOREM 5.10.

$$dev \ G = \frac{1}{\rho^4 n^4}\sum_{x,y}(|N_x \cap N_y| - \rho^2 n)^2$$

where $N_x = N(x) = \{y : y \sim x\}$ and $d_x = |N_x|$.

PROOF.

$$\rho^4 n^4 dev \ G = \sum_{x,y,z,w}\chi(x,z)\chi(y,z)\chi(x,w)\chi(z,w)$$

$$= \sum_{x,y}(\sum_z \chi(x,z)\chi(y,z))^2$$

$$= \sum_{x,y}[|N_x \cap N_y|(1 - \rho)^2 - (|N_x \cap \bar{N}_y| + |\bar{N}_x \cap N_y|)\rho(1 - \rho)$$

$$+ |\bar{N}_x \cap \bar{N}_y|\rho^2]$$

For fixed x, y, we have

$$|N_x \cap N_y|(1 - \rho)^2 - |N_x \cap \bar{N}_y|\rho(1 - \rho) - |\bar{N}_x \cap N_y|\rho(1 - \rho) + |\bar{N}_x \cap \bar{N}_y|\rho^2$$

$$= (|N_x \cap N_y| - \rho^2 n)^2 - (|N_x| - \rho n)\rho - (|N_y| - \rho n)\rho.$$

Therefore we have

$$\rho^4 n^4 dev \ G = \sum_{x,y}(|N_x \cap N_y| - \rho^2 n)^2$$

and Theorem 5.10 is proved. $\qquad\square$

The above Theorem is useful for relating the neighborhood property (e.g., P_4) to the deviation property.

THEOREM 5.11. *For a regular graph G, we have*

$$dev \ G = \sum_{i \neq 0}(1 - \lambda_i)^4$$

PROOF. We consider the matrix

$$M = I - \mathcal{L} - \frac{1}{\text{vol } G}T^{1/2}J(T^{1/2})^*$$

whose eigenvalues are exactly 0 and $1 - \lambda_i$ for $i = 1, \ldots, n - 1$. Considering the trace of M^4, we obtain

$$(5.4) \qquad \sum_{i \neq 0} (1 - \lambda_i)^4 = tr(M^4).$$

On the other hand, it is easy to check that

$$M(x, y) = \frac{\chi(x, y)}{\rho n}.$$

Hence,

$$\begin{aligned} tr(M^4) &= \sum_{x,y,z,w} \frac{\chi(x, z)\chi(y, z)\chi(x, w)\chi(y, w)}{\rho^4 n^4} \\ &= dev \ G \end{aligned}$$

$\square$

THEOREM 5.12. *For any graph G, we have*

$$dev \ G \ \leq \ \sum_{i \neq 0} (1 - \lambda_i)^4 + 20\sqrt{irr \ G}.$$

The proof is essentially the same as the proof above, but somewhat messier. We will skip the proof here.

So, we see that small $\bar{\lambda}$ and the condition of being almost regular together imply that the deviation of G is small. From Theorem 5.10, small deviation in turn implies that for almost all x, y, $N_x \cap N_y$ is close to the expected size $\rho^2 n$. Since the count of (labelled) 4-cycles as subgraphs is exactly $\sum_{x,y} |N_x \cap N_y|^2$, small $dev \ G$ implies that this count is close to the expected number of such subgraphs in a random graph with the same edge density. While the deviation is very effective in dealing with graphs with large edge density, we point out here that for sparse graphs (for example, graphs with no 4-cycles), the deviation is not as useful. Still, for dense graphs, the deviation is quite powerful. For example, all of the quasi-random properties in the list in Section 5.1 are implied by the deviation property since, in fact, $dev \ G = o(1)$ can also be added to the list of (equivalent) quasi-random properties. The interrelations of various quasi-random properties will be discussed in the next section.

5.4. Quasi-random graphs

To simplify the discussion, we assume that all graphs G in this section are almost regular, i.e.,

(P_r)

$$irr \ G = \frac{\rho}{n} \sum_{v \in V(G)} \left(\frac{1}{\rho_v} - \frac{1}{\rho} \right) = o(1)$$

where $\rho_v = d_v/n$, $\rho = \sum d_v/n^2$ and $|V(G)| = n$. Our goal here is to derive equivalence relations between various graph invariants and the eigenvalue property:

(P_e)

$$\bar{\lambda} = \max_{i \neq 0} |1 - \lambda_i| = o(1)$$

where λ_i's are eigenvalues of the Laplacian of G.

Some directions of these relations have already been proved in the previous sections. Here we first consider bounding the discrepancy of a graph by $\bar{\lambda}$ by using Theorem 5.1.

THEOREM 5.13. *For a graph G, the discrepancy of G satisfies*

$$disc\ G \leq (\bar{\lambda} + 2\sqrt{irr\ G})\mathrm{vol}\ G$$

where $\bar{\lambda} = \max_{i \neq 0} |1 - \lambda_i|$.

To prove Theorem 5.13, we need to say a little more about quantities associated with being almost regular. As we recall, the irregularity $irr\ G$ of G was defined in connection with the proofs involving deviation. A simpler definition could be the variance

$$var\ G = \frac{1}{\rho^2 n} \sum_v (\rho_v - \rho)^2.$$

Ostensibly, the variance $var\ G$ goes to zero if and only if $irr\ G$ does. It would simplify our discussion if these two quantities were "equivalent" in the sense that "$var\ G \leq c_1\ irr\ G$" and "$irr\ G \leq c_2\ var\ G$" for some constants c_1 and c_2. However only the first of these two relations is true (S_n, a star on n vertices serves as a example.) In addition, the statement of Theorem 5.12 will not hold if we replace $irr\ G$ by $var\ G$. This dictates then our choice of taking the upper bound as $irr\ G$.

LEMMA 5.14. *For positive reals $a_1, \ldots, a_n$ with average a, we have*

$$\frac{1}{a^2 n^2} \sum_{i=1}^n (a_i - a)^2 \leq \frac{a}{n} \sum_{i=1}^n \left(\frac{1}{a_i} - \frac{1}{a} \right).$$

The proof is straightforward. Consequently, we have

$$var\ G \leq irr\ G.$$

In fact, for any subset X of $V(G)$, we have

$$(5.5) \qquad var\ G \quad \geq \quad \frac{1}{\rho^2 n} \sum_{v \in X} (\rho_v - \rho)^2$$

$$\geq \quad \frac{1}{\rho^2 n \mid X \mid} \left(\sum_{v \in X} (\rho_v - \rho) \right)^2$$

$$= \quad \frac{1}{\rho^2 n^3 \mid X \mid} (\mathrm{vol}\ X - \rho n |X|)^2$$

Before we proceed to prove Theorem 5.13, we remark that there are two possible ways that we could define discrepancy:

$$\sup_{X, Y \subseteq V(G)} |e(X, Y) - \rho |X|\ |Y|\ |$$

or

$$\sup_{X \subseteq V(G)} |e(X,X) - \rho|X|^2 |.$$

These two quantities are bounded from each other to within a constant factor since

$$e(X \cup Y, X \cup Y) = e(X,X) + e(Y,Y) + 2e(X,Y).$$

So we will use whichever version that is convenient for the proof, always allowing for a constant factor.

Proof of Theorem 5.13:
From Theorem 5.2, we have

$$|e(X,X) - \frac{(\mathrm{vol}\ X)^2}{\mathrm{vol}\ G}| \le \bar{\lambda}\mathrm{vol}\ X.$$

Therefore,

$$
\begin{aligned}
disc\ G &= \sup_{X \subseteq V(G)} |e(X,X) - \rho|X|^2| \\
&\le \sup_{X \subseteq V(G)} |e(X,X) - \frac{(vol\ X)^2}{\mathrm{vol}\ G}| + |\frac{(\mathrm{vol}\ X)^2}{\mathrm{vol}\ G} - \rho|X|^2| \\
&\le \bar{\lambda}\mathrm{vol}\ X + |\frac{(\mathrm{vol}\ X)^2}{\mathrm{vol}\ G} - \rho|X|^2|.
\end{aligned}
$$

From Lemma 5.7 we have

$$
\begin{aligned}
\frac{(\mathrm{vol}\ X)^2}{\rho n^2} - \rho|X|^2 &= \frac{(\mathrm{vol}\ X - \rho n|X|)}{\rho n^2}(\mathrm{vol}\ X + \rho n|X|) \\
&\le 2(\mathrm{vol}\ X - \rho n|X|) \\
&\le 2\sqrt{var\ G} \cdot \rho n^2 \\
&\le 2\sqrt{irr\ G} \cdot \rho n^2.
\end{aligned}
$$

The proof of Theorem 5.13 is complete. $\square$

We remark that we can replace *irr G* by *var G* in the statement of Theorem 5.13 by using (5.5).

Theorem 5.13 indicates that the eigenvalue property P_e and the almost regular property imply the discrepancy property:

(P_{disc}) $disc\ G = o(\mathrm{vol}\ G).$

In other words, $P_e + P_r \Rightarrow P_{disc}$.

We will denote the deviation property as follows:

(P_{dev}) $dev\ G = o(1).$

From Theorem 5.12, we have

$$|dev\ G - \sum_{i \ne 0}(1 - \lambda_i)^4| \le 20\sqrt{irr\ G}.$$

As a consequence of the above inequalities and the fact that $\bar{\lambda}^4 \le \sum_{i \ne 0}(1 - \lambda_i)^4$, we have $P_{dev} + P_r \Rightarrow P_e$. To complete the equivalence of the three properties $P_e + P_r$, $P_{disc} + P_r$, $P_{dev} + P_r$, we will prove

THEOREM 5.15. *For a graph G, we have*

$$disc\ G \le \rho n^2 (dev\ G)^{1/4},$$

$$dev\ G \le \frac{4}{\rho^4 n^2} disc\ G + 2\ var\ G \le \frac{4}{\rho^4 n^2} disc\ G + 2\ irr\ G.$$

PROOF. First we consider, for $X, Y \subset V$,

$$
\begin{aligned}
\rho^4 n^4 dev\ G &= \sum_{x,y,z,w} \chi(x,y)\chi(y,z)\chi(z,w)\chi(w,x) \\
&= \sum_{x,z}(\sum_y \chi(x,y)\chi(y,z))^2 \\
&\ge \sum_{x,z \in X}(\sum_y \chi(x,y)\chi(y,z))^2 \\
&\ge \frac{1}{|X|^2}(\sum_{x,z \in X}\sum_y \chi(x,y)\chi(y,z))^2 \\
&\ge \frac{1}{|X|^2}(\sum_y(\sum_{x \in X}\chi(x,y))^2)^2 \\
&\ge \frac{1}{|X|^2}(\sum_{y \in Y}(\sum_{x \in X}\chi(x,y))^2)^2 \\
&\ge \frac{1}{|X|^2|Y|^2}(\sum_{y \in Y}\sum_{x \in X}\chi(x,y))^4 \\
&= \frac{1}{|X|^2|Y|^2}(e(X,Y) - \rho|X|\,|Y|)^4
\end{aligned}
$$

This implies

$$disc\ G \le \rho n^2 (dev\ G)^{1/4},$$

For the second inequality, we consider the deviation of G using Theorem 5.10:

$$\rho^4 n^4 dev\ G = \sum_{x,y}(|N_x \cap N_y| - \rho^2 n)^2.$$

For a fixed x, we consider

$$W_x = \sum_y (|N_x \cap N_y| - \rho^2 n)^2.$$

We note that $|N_x \cap N_y|$ is exactly the number $d'_y = |e(y, N_x)|$ of edges from y to the set N_x. We consider the set X consisting of the vertices y with $d'_y \le \rho d_x$. We have

$$\sum_{y \in X}|d'_y - \rho d_x| = |\ e(N_x, X) - \rho d_x|X|\ | \le disc\ G.$$

Similarly,

$$\sum_{z \notin X} |d'_z - \rho d_x| = | \, e(N_x, \bar{X}) - \rho d_x |\bar{X}| \, | \leq disc \ G$$

Therefore

$$\sum_y (d'_y - d_x \rho)^2 \leq 2n \ disc \ G$$

and

$$
\begin{aligned}
W_x &= \sum_y (|N_x \cap N_y| - \rho^2 n)^2 \\
&\leq \sum_y 2(|N_x \cap N_y| - \rho^2 n)^2 + 2n(\rho d_x - \rho^2 n)^2 \\
&\leq 4n \ disc \ G + 2\rho^2 n^3 (\rho_x - \rho)^2.
\end{aligned}
$$

Hence

$$\rho^4 n^4 dev \ G \ \leq \ \sum_x W_x \leq 4n^2 disc \ G + 2\rho^2 n^2 var \ G.$$

$\square$

We have so far proved the equivalence of the discrepancy, deviation, and eigen-value properties for almost regular graphs. In fact, all of the properties P_1, P_2, P_3, P_4 have been covered for graphs with edge density $1/2$. As for the property $P_5(s)$, for any graph $M(s)$ on s vertices, $s \geq 4$, the number $N^*_G(M(s))$ of occurrences of labelled induced subgraphs of G isomorphic to $M(s)$ satisfies

$$N^*_G(M(s)) = (1 + o(1))n^s 2^{-\binom{s}{2}}.$$

The proof of the equivalence of $P_5(s)$, for $s \geq 4$, with the other quasi-random properties is somewhat complicated and can be found in [76]. A quantitative version was proved in [74]. The difference of $N^*_G(M(s))$ and its expected value can be bounded by a function of the deviation of G. Namely, for a graph G with edge density $1/2$, we have

$$|N^*_G(M(s)) - n^s 2^{-\binom{s}{2}}| \leq n^s (dev \ G)^{1/4}.$$

This property is quite strong in the sense that quasi-randomness implies that if a graph G with edge density $1/2$ contains $(1 + o(1))\frac{n^4}{16}$ C_4's as labelled subgraphs, then G contains the expected number of Petersen graphs (as well as any graphs of a fixed size). On the other hand, there are examples of graphs containing the correct number of all subgraphs on 3 vertices but which nonetheless fail to be quasi-random.

Notes:
Quasi-randomness resides at the intersection of extremal graph theory and random graph theory. The original paper [76] (also see [72, 227]) on quasi-random graphs provides a general framework for examining the relationships of combinatorial structures as well. There are several papers on quasi-random hypergraphs [67, 70, 74, 77] and their connection with cohomology [78] and communication complexity [88]. Work has also been done on quasi-random sequences, subsets

[**75**] and tournaments [**71**]. Many questions on the quantitative analysis of the $o(\cdot)$ terms remain open.

CHAPTER 6

Expanders and explicit constructions

6.1. Probabilistic methods versus explicit constructions

Many problems in combinatorics, theoretical computer science and communication theory can be solved by the following probabilistic approach: To prove the existence of some desired object, first an appropriate (probability) measure is defined on the class of objects; second, the subclass of desired objects in question are shown to have positive measure. This implies that the desired objects must exist. This technique, while extremely powerful, suffers from a serious drawback. Namely, it gives no information about how one might actually go about explicitly constructing the desired objects. Thus, while we might even be able to conclude that almost all of our objects have the desired property (that is, all except for a set of measure zero), we may be unable to exhibit a single one. A simple example of this phenomenon from number theory is that of a normal number. A real number x is said to be *normal* if for each integer $b \geq 2$, each of the digits $0, 1, \cdots, b-1$ occurs asymptotically equally often in the base b expansion of x. It is known that almost all (in Lebesgue measure) real numbers are normal, but no one has yet succeeded in proving that any particular number (such as π, e or $\sqrt{2}$) is normal.

One of the earliest examples of the above probabilistic method is Erdős' classical result [**111**] from 1947 on the existence of graphs on n vertices which have maximum cliques and independent sets of size $2 \log n$ (see Chapter 5, Section 2). Since then, probabilistic methods have been successfully used in a wide range of areas. However, in spite of the success of probabilistic methods, there is a clear need for explicit constructions, especially for applications in algorithmic design and in building efficient communication networks.

In the past ten years, substantial progress has been made on explicit constructions of so-called expander graphs. Although the developments in constructing expander graphs were first motivated by applications in communication networks, it has become one of the most powerful tools in complexity theory, parallel architectures, distributed computing, derandomized algorithm, and the list goes on [**30, 120, 208**].

Still, we have not yet rigorously defined "explicit constructions" versus "random graphs." So, some explanations are in order. When we mention random graphs, some probability distributions are involved. For example, a typical model for graphs on n vertices is to assign equal probability to each of the $2^{\binom{n}{2}}$ possible graphs on n (labelled) vertices. We say a random graph G on n vertices satisfies property P if all except for $o(1)$ of the graphs satisfy P where the $o(1)$ term goes to zero as

n approaches infinity. On the other hand, by an explicit construction, we mean a deterministic description from which we can construct a graph on n vertices with the number of steps polynomial in n, for infinitely many n. In other words, an explicit construction is a polynomial time algorithm for defining a graph. In fact, all of the explicit constructions we describe have very short descriptions. This, of course, is just one possible interpretation of "explicit construction".

What is an expander graph? Roughly speaking, in an expander graph, any "small" subset of vertices has a relatively "large" neighborhood. We will leave the detailed description of the quantitative definitions of "small" and "large" until the next section. Conceivably, an expander graph can allow us to build networks with guaranteed access for making connections or routing messages.

The key to the success in constructing expander graphs is its relationship with eigenvalues. Such a relationship can be used in two ways.

1. Validating a random construction.
One of the shortfalls of the probabilistic method is that it only guarantees the existence of an expander. A random graph, which is usually generated by using a random source (e.g. flipping a coin, etc.), has high probability being an expander. Since eigenvalues can be computed easily (in polynomial time), we can validate the randomly generated graph by computing its eigenvalues and the associated expander constants. While the probabilistic method does not guarantee the generation of expanders, it assures us that with high probability (say, within a few tries) we can obtain graphs whose performance in expansion is backed up by the eigenvalue bounds.

2. Constructing large families of expanders.
A general eigenvalue lower bound can often be used to establish a whole family of expander graphs. In particular, algebraic and number-theoretic techniques can be used to bound the eigenvalues for special (infinite) families of graphs.

In the next section we will discuss various notions of expansion properties and their relations to eigenvalues. A variety of examples and explicit constructions will be given including Paley graphs, coset graphs, Margulis graphs, and Ramanujan graphs. In Section 6.4, we describe some classical applications of expanders to communication networks and we illustrate how to use expander graphs to create various building blocks for these networks. In Section 6.5, we discuss constructions for graphs with small diameter and girth. In the last section, we consider an extremal invariant of a graph which depends on the maximum eigenvalues of all possible weighted Laplacians. This invariant is closely related to Lovasz's ϑ function and it can be computed in polynomial time. We will describe its relationship to the chromatic number and clique number of a graph.

6.2. The expanders

The expansion property is crucial in many applications [**24, 175, 183, 211, 212, 209, 240, 245**] and has become a driving force for recent progress in constructive methods. Success here is due, in large part, to a combination of tools from graph theory, network theory, theoretical computer science and various mathematical areas such as number theory, representation theory, and harmonic analysis. Perhaps, because of the large number of different applications in disparate settings, the definitions of expansion-like properties vary from one situation to another, often with a variety of names such as expander, magnifier, enlarger, generalizer, concentrator, and superconcentrator, just to name a few. To make matters worse, most of these definitions involve a large number of parameters. One typical example, namely the definition of a concentrator, is as follows: An $(n, \theta, k, \alpha, \beta)$-concentrator is a bipartite graph with n inputs, θn outputs and $k\,n$ edges, such that every input subset A with $\mid A \mid \geq \alpha n$ has at least βn neighbors. It is conceivable that such tedious definitions actually hindered early progress in this area.

The expansion property basically means each subset X of vertices must have "many" neighbors. That is, the neighborhood set $N(X) = \{y : y$ is adjacent to some $x \in X\}$ is "large" in comparison to the size of X. The difficulty lies in finding a good way to define "many" or "large," quantitatively. There is an obvious scenario in which the subset S is almost the entire vertex set and the vertex boundary $\delta(S) = N(S) - S$ is very small. The typical definition of expander graphs is as follows: A regular graph G on m vertices is a c-expander if every subset S of $V(G)$ satisfies

$$\mid \delta(S) \mid \geq c \left(1 - \frac{\mid S \mid}{n} \right) \mid S \mid \ .$$

Here the constant c is often called the *expander coefficient*.

Another way to define the expansion factor is as follows:

$$c' = \sup_{S \subseteq V(G)} \frac{|\delta(S)|}{|S| \cdot |\bar{S}|/n}.$$

We remark that this definition is somewhat unsatisfactory for certain problems since the expansion factor c' and the degree k are intimately related. For example, a random regular graph of degree k has an expansion factor of roughly k when the subset is small. The values of c and c' should be judged in comparison with a function of k. In such cases, it works out better to formulate the lower bound for $|\delta(S)|$ directly in terms of eigenvalues.

The expansion factor c' is closely related to the Cheeger constant. In particular, for a regular graph G,

$$c' \geq g_G \geq \frac{1}{2} c'$$

where $g_G := \inf_S \dfrac{\text{vol } \delta S}{\min(\text{vol } S, \text{vol } \bar{S})}$, as defined in Chapter 2.

The expansion of G is closely related to the discrepancy of G in the following sense: The discrepancy property implies every subset S contains close to the expected number of edges; therefore there are "many" edges leaving S.

Another related invariant is the *isoperimetric number* [**45, 200**], denoted by $i(G)$, and defined by

$$i(G) = \min_{S \subset N, |S| < \frac{n}{2}} \frac{|\{\{u,v\} \in E(G) : u \in S, v \notin S\}|}{|S|}.$$

The so-called *conductance* is $1/k$ times $i(G)$ for a k-regular graph G [**167**]. Clearly, for a k-regular graph, we have also $i(G) \geq k/2 - 2\,disc(G)$.

We will give simple proofs which establish eigenvalue upper bounds for the expansion coefficient. Although the problem of checking whether a graph is an (n, k, c)-expander is *co-NP-complete* [**29**],the following relationship provides an efficient method to estimate the expansion and discrepancy of a graph. In Section 3.2, a similar bound was derived by examining the relation between diameter and eigenvalues.

LEMMA 6.1. *For a graph G and $S \subseteq V(G)$, the vertex boundary $\delta(S) = \{x \notin S : x \sim y \in S\}$ satisfies*

$$\frac{vol\,\delta(S)}{vol\,S} \geq \frac{\lambda(2-\lambda)\dfrac{vol\,\bar{S}}{vol\,G}}{1 - \lambda(2-\lambda)\dfrac{vol\,\bar{S}}{vol\,G}}$$

$$\geq \lambda(2-\lambda)\frac{vol\,\bar{S}}{vol\,G}$$

where $\lambda = 2\lambda_1/(\lambda_{n-1} + \lambda_1)$. In other words, G is a $\lambda(2-\lambda)$-expander.

PROOF. It follows from Theorem 3.1 and Lemma 3.4 of Chapter 3 that

$$\frac{vol\,\delta S}{vol\,S} \geq \frac{1 - (1-\lambda)^2}{(1-\lambda)^2 + vol\,S/vol\,\bar{S}}.$$

By straightforward calculation, we get

$$\frac{vol\,\delta(S)}{vol\,S} \geq \frac{(2\lambda - \lambda^2)vol\,\bar{S}}{(1-\lambda)^2 vol\,\bar{S} + vol\,S}$$

$$= \frac{\lambda(2-\lambda)\dfrac{vol\,\bar{S}}{vol\,G}}{1 - \lambda(2-\lambda)\dfrac{vol\,\bar{S}}{vol\,G}}$$

$$\geq \lambda(2-\lambda)\frac{vol\,\bar{S}}{vol\,G}$$

as desired.   □

For a graph G and $S \subseteq V(G)$, we recall that the neighborhood $N(S)$ of a subset S is defined as follows:

$$N(S) = \{x : x \sim y \in S\}.$$

Note that S is not necessarily contained in $N(S)$. We consider the following variations of the expander lower bounds:

LEMMA 6.2. *Suppose G is not a complete graph. For $S \subseteq V(G)$, the neighborhood $N(S)$ satisfies*

$$\frac{\text{vol } N(S)}{\text{vol } S} > \frac{1}{\bar{\lambda}^2 + (1 - \bar{\lambda}^2)\frac{\text{vol } S}{\text{vol } G}} = \frac{1}{1 - (1 - \bar{\lambda}^2)\frac{\text{vol } \bar{S}}{\text{vol } G}}$$

where $\bar{\lambda} = \max_{i \neq 0} |1 - \lambda_i|$.

PROOF. In the proof of Theorem 3.1, suppose that we choose $p_t(\mathcal{L})$ to be $I - \mathcal{L}$ and $Y = V(G) - N(X)$. Then we have

$$\begin{aligned}
0 &= \langle T^{1/2}\psi_Y, p_t(\mathcal{L})T^{1/2}\psi_X \rangle \\
&> \frac{\text{vol } X \text{ vol } Y}{\text{vol } G} - \bar{\lambda}\frac{\sqrt{\text{vol } X \text{ vol } \bar{X} \text{ vol } Y \text{ vol } \bar{Y}}}{\text{vol } G}.
\end{aligned}$$

Thus

$$\bar{\lambda}^2\text{vol } \bar{X} \text{ vol } \bar{Y} > \text{vol } X \text{ vol } Y.$$

For $\bar{Y} = N(X)$, this implies

$$\frac{\text{vol } N(S)}{\text{vol } S} > \frac{1}{\bar{\lambda}^2 + (1 - \bar{\lambda}^2)\frac{\text{vol } S}{\text{vol } G}} = \frac{1}{1 - (1 - \bar{\lambda}^2)\frac{\text{vol } \bar{S}}{\text{vol } G}}.$$

$\square$

For regular graphs, we give a direct proof here.

LEMMA 6.3. *Suppose G is a regular graph on n vertices and G is not a complete graph. For $S \subseteq V(G)$, the neighborhood $N(S)$ satisfies*

$$\frac{|N(S)|}{|S|} > \frac{1}{\bar{\lambda}^2 + (1 - \bar{\lambda}^2)\frac{|S|}{n}} = \frac{1}{1 - (1 - \bar{\lambda}^2)\frac{|\bar{S}|}{n}}$$

where $\bar{\lambda} = \max_{i \neq 0} |1 - \lambda_i|$.

PROOF. For $S \subset V(G)$, we consider the characteristic function ψ_S:

$$\psi_S(x) = \begin{cases} 1 & \text{if } x \in S \\ 0 & \text{otherwise} \end{cases}$$

We consider the following inner product:

$$\begin{aligned}
\langle \psi_S A, A\psi_S \rangle / k^2 &= \sum a_i^2(1 - \lambda_i)^2 \\
&< a_0^2 + (\sum_{i \geq 1} a_i^2)\bar{\lambda}^2 \\
(6.1) \qquad &\leq (1 - \bar{\lambda}^2)\frac{|S^2|}{n} + \bar{\lambda}^2|S|.
\end{aligned}$$

On the other hand,

$$\langle \psi_S A, A\psi_S \rangle = \sum_{u \in S} \sum_{v \in S} |\{w : \{v, w\} \in E \text{ and } \{u, w\} \in E\}|$$

$$= \sum_{w \in V} |N(w) \cap S|^2 .$$

Applying the Cauchy-Schwarz inequality, we have:

$$\sum_{w \in V} |N(w) \cap S|^2 \geq \frac{(\sum_{w \in V} |N(w) \cap S|)^2}{|N(S)|}$$

$$= \frac{|S^2|k^2}{|N(S)|}.$$

Combining this with (6.1), we obtain

$$(6.2) \qquad\qquad |N(S)| > \frac{|S|}{(1 - \rho^2)\frac{|S|}{n} + \rho^2}.$$

$\square$

For a bipartite graph, we can have a modified version for the expander theorem. For a bipartite graph G with vertex set $X \cup Y$ and edges between X and Y, the incidence matrix $M = M(G)$ has columns indexed by vertices in X and rows indexed by vertices in Y. For $x \in X$ and $y \in Y$, the matrix M satisfies $M(x, y) = 1$ if and only if $\{x, y\}$ is an edge. A bipartite expander graph depends on the eigenvalues of M^*M as follows:

LEMMA 6.4. *For a bipartite graph G with vertex set $X \cup Y$ and edges between X and Y, suppose all vertices in X have the same degree. For a subset S of X, the neighborhood $N(S)$ satisfies:*

$$\frac{|N(S)|}{|S|} \geq \frac{1}{\rho^2 + (1 - \rho^2)\frac{|S|}{n}} = \frac{n}{|S| + (1 - \rho^2)|\bar{S}|}$$

where $|X| = n$ and $\rho^2 k^2$ is the second largest eigenvalue of M^*M where $M = M(G)$ is the bipartite adjacency matrix defined above.

PROOF. Suppose every vertex in X has degree k. Let ρ_i^2 denote the eigenvalues of M^*M. We denote the largest eigenvalue of M^*M by $\rho_0^2 = k^2$. Let the a_i's denote the Fourier coefficients of characteristic function ψ_S with respect to the eigenfunctions of M^*M. We consider the following inner product:

$$(6.3) \qquad\qquad \frac{1}{k^2}\langle \psi_S M^*, M\psi_S \rangle = \sum a_i^2 \rho_i^2$$

$$\leq (1 - \rho^2)\frac{|S^2|}{n} + \rho^2|S|.$$

Together with

$$\langle \psi_S M^*, M\psi_S \rangle \;=\; \sum_{u \in S}\sum_{v \in S} |\,\{w : \{v,w\} \in E \text{ and } \{u,w\} \in E\}\,|$$

$$= \sum_{w \in Y} |\,N(w) \cap S\,|^2$$

$$\geq \frac{(\sum_{w \in V} |\,N(w) \cap S\,|)^2}{|\,N(S)\,|}$$

$$= \frac{|S|^2 k^2}{|\,N(S)\,|}.$$

Combining this with (6.3), we have

$$|\,N(S)\,| \geq \frac{|S|}{(1 - \rho^2)\frac{|S|}{n} + \rho^2}.$$

$\square$

We remark that if we consider the following modified matrix $\mathcal{M}$ for a bipartite graph (not necessarily regular), then the above proof works in a similar way. We define

$$\mathcal{M}(u,v) = \frac{1}{\sqrt{d_y}}$$

LEMMA 6.5. *For a bipartite graph G with vertex set $X \cup Y$ and edges between X and Y, and for a subset S of X, the neighborhood $N(S)$ satisfies:*

$$\frac{|N(S)|}{|S|} \geq \frac{1}{\rho^2 + (1 - \rho^2)\frac{|S|}{n}} = \frac{n}{|S| + (1 - \rho^2)|\bar{S}|}$$

where $|X| = n$ and ρ^2 is the second largest eigenvalue of $\mathcal{M}^*\mathcal{M}$.

We remark that the inequalities in Lemmas 6.4 and 6.5 are strict if the eigenvalues except for the largest are not all equal.

6.3. Examples of explicit constructions

We will describe several useful families of expander graphs. For each construction, the precise value of a bound for the eigenvalues will be discussed. Together with the theorems in Section 6.2, these constructions yield good expanders. We will begin with graphs with edge density about $\frac{1}{2}$ and then proceed to graphs with lower edge density, say $\frac{k}{n}$ for fixed k.

Construction 6.3.1. The Paley graph P_p.

Let p be a prime number congruent to 1 modulo 4. The Paley graph consists of p vertices, $0, 1, 2, \cdots, p-1$. Two vertices i and j and adjacent if and only if $i - j$ is a non-zero quadratic residue modulo p. The eigenvalues of P_p are exactly

$$1 - \sum_{x \in Z_p^*} e^{\frac{2\pi i j x^2}{p}}/(p-1)$$

for each $j = 0, \cdots, p - 1$. These are closely related to Gauss sums modulo p (see [166]). In particular, it is known that for any $j \not\equiv 0 \pmod{p}$, the Gauss sum $\sum_{x \in Z_p^*} e^{\frac{2\pi i j x^2}{p}}$ is either $\sqrt{p} - 1$ or $-\sqrt{p} - 1$.

Therefore, we have

$$\lambda_1 = 1 - \frac{\sqrt{p} - 1}{2(p-1)},$$

$$\lambda_{p-1} = 1 + \frac{\sqrt{p} + 1}{2(p-1)}.$$

The Paley graph is a favorite and frequently cited example in extremal graph theory, no doubt due to its many nice properties the existence of which are guaranteed by eigenvalue bounds. For example, P_p contains all induced subgraphs on $c\sqrt{\log p}$ vertices (see [35]; also implicit in [146]).

Construction 6.3.2. The Paley sum graphs $\tilde{P}_p$.

The Paley sum graphs are basically the symmetric versions of Paley graphs without the constraint $p \equiv 1 \pmod 4$. Let p be any prime number. $\tilde{P}_p$ has vertices $0, \cdots, p - 1$, and two vertices i and j are adjacent if and only if $i + j$ is a quadratic residue modulo p. The eigenvectors for Paley graphs are ϕ_j where $\phi_j(k) = e^{2\pi i j k}$, for $j = 0, \ldots, n - 1$. For sum graphs, the eigenvectors are

$$\phi_j + \frac{(1 - \lambda_j)}{\sqrt{(1 - \lambda_j)(1 - \lambda_{-j})}} \phi_{-j}$$

where λ_j is the eigenvalue for the eigenvector ϕ_j of the Paley graph. Therefore the eigenvalues for the sum graphs are exactly $1 \pm \sqrt{|(1 - \lambda_j)(1 - \lambda_{-j})|}$. For $p \equiv 1 \pmod 4$, the eigenvalues of the Paley sum graph is the same as those of the Paley graph. For $p \equiv 3 \pmod 4$, the eigenvalues of the Paley sum graph are 0, $1 - \frac{1}{2\sqrt{p-1}}$ (with multiplicity $\frac{p-1}{2}$) and $1 + \frac{1}{2\sqrt{p-1}}$ (with multiplicity $\frac{p-1}{2}$).

The Paley graphs and Paley sum graphs both have edge density about $\frac{1}{2}$. They can be generalized to graphs with edge density $\frac{t}{r}$ for any fixed constants t and r with $t < r$. Paley sum graphs are actually a special case of the following:

Construction 6.3.3. The generalized Paley sum graphs $P_{p,r,T}$.

For a fixed integer $r > 0$, let $p = mr + 1$ be a prime congruent to 1 mod 4 and let $T \subset \mathcal{Z}_p^*$ consist of t non-zero residues so that for any distinct $a, b \in T$, ab^{-1} is not an rth power in $\mathcal{Z}_p^*$. The generalized Paley graph has vertex set $\{0, 1, \cdots, p - 1\}$. Two vertices i and j are adjacent if and only if $i + j = aq$ for some $a \in T$ and q an rth power. The eigenvalues are $\lambda_j = (1 - \sum_{a \in T} \sum_{x \in Z_p^*} e^{2\pi i j a x^2})/(p - 1)t$.

For $j \neq 0$, using the well-known theorem of Deligne [94], we have

$$|\sum_{x \in Z_p} e^{2\pi i j x^2}| \leq (r - 1)\sqrt{p}.$$

Therefore the eigenvalues of the generalized Paley sum graphs $P_{p,r,t}$ satisfy

$$\lambda_1 \geq 1 - \frac{(r-1)\sqrt{p}+1}{p-1}, \ \lambda_{n-1} \leq 1 + \frac{(r-1)\sqrt{p}+1}{p-1}.$$

In the other direction, Paley graphs can be generalized to the following coset graphs on n vertices with edge density $n^{-1+\frac{1}{t}}$ for any positive integer t (see [51]).

Construction 6.3.4. The coset graphs $C_{p,t}$.

We consider the finite field $GF(p^t)$ and a coset $x + GF(p)$ for $x \in GF(p^t) \simeq GF(p)(x)$. There is a natural correspondence between elements of the multiplicative group $GF^*(p^t)$ and elements of $\{1, \cdots, p^t - 1\}$. For example, choosing a generator g, each element y in $GF^*(p^t)$ corresponds to the integer k for which $y = g^k$. Now we consider the coset graph $C_{p,t}$ with vertices $1, \cdots, p^t - 1 = n$, and edges $\{a, b\}$ if $a + b$ is in the subset X of integers corresponding to the coset $x + GF(p)$. The eigenvalues of the coset graph $C_{p,t}$ are $\sum_{a \in X} \theta^a$ for θ ranging over all nth roots of 1.

Bounding the eigenvalues of coset graphs leads to a natural generalization of Weil's character sum inequality [166]. The following inequality was conjectured by the author [51] and proved by Katz [172] and others [184, 186]. If θ is a $(p^t - 1)$-th root of 1 and $\theta \neq 1$, we have

$$\left| \sum_{a \in X} \theta^a \right| \leq (t-1)\sqrt{p},$$

where X is the coset $x + GF(p)$.

The coset graph has edge density $n^{1 - \frac{1}{t}}$, and the eigenvalues satisfy $\lambda_1 \geq 1 - \frac{t-1}{\sqrt{p}}$ and $\lambda_{n-1} \leq 1 + \frac{t-1}{\sqrt{p}}$.

Construction 6.3.5. The Margulis graphs M_n.

In the early 70's, Margulis [195] ignited an entire movement toward the study of constructive methods by relating Kazhdan's property T to expanders. This approach was successfully continued by Gabber and Galil [137] who obtained explicit values for estimating the expander constant. Here we illustrate some of these early constructions of these very elegant graphs, which we call Margulis graphs [8, 137, 195]. Set $n = m^2$ and $V = Z_m \times Z_m$. Consider the following six transformations from V to itself:

$$\begin{aligned}
\sigma_1(x,y) &= (x, y+2x) \\
\sigma_2(x,y) &= (x, y+2x+1) \\
\sigma_3(x,y) &= (x, y+2x+2) \\
\sigma_4(x,y) &= (x+2y, y) \\
\sigma_5(x,y) &= (x+2y+1, y) \\
\sigma_6(x,y) &= (x+2y+2, y)
\end{aligned}$$

(Addition here is modulo m.)

Let $G = M_n = (V, E)$ be the graph with vertex set V and with edges $\{u, v\}$ if $u = \sigma_i(v)$ or $v = \sigma_i(u)$ for some i. (Thus, e.g., $(0,0)$ is joined to itself by 2 loops -

note that here we consider that a loop adds 2 to the degree of a vertex.) Obviously, G is 12-regular.

Claim:

$$\lambda_1 \geq \frac{2 - \sqrt{3}}{3}.$$

PROOF. Let T be the $(0,1) \times (0,1)$ torus, and define two measure-preserving automorphisms ψ_1, ψ_2 on T by $\psi_1(x,y) = (x, y + 2x), \psi_2(x,y) = (x + 2y, y)$, where addition is modulo 1.

The main part of the proof is based on the following fact proved as Lemma 4 of [137]. Here we state without proof this analytic result:

If ϕ is measurable on T and $\int_T \phi = 0$, then

(6.4)
$$\int_T |\phi \cdot \psi_1^{-1} - \phi|^2 + \int_T |\phi \cdot \psi_2^{-1} - \phi|^2 \geq c \int_T \phi^2,$$

where $c = 4 - \sqrt{12}$.

Now suppose that $f : V \to \mathbb{R}$ satisfies $\sum_{j,k=1}^m f(j,k) = 0$. Define a measurable function $\phi : T \to \mathbb{R}$ as follows: If $(j,k) \in Z_m \times Z_m$ then for

$$\frac{j}{m} \leq x < \frac{j+1}{m}, \quad \frac{k}{m} \leq y < \frac{k+1}{m}, \quad \phi(x,y) = f(j,k).$$

It can be checked that $\int_T \phi = 0$,

and

$$\int_T |\phi \cdot \psi_1^{-1} - \phi|^2 + \int_T |\phi \cdot \psi_2^{-1} - \phi|^2$$

$$= \frac{1}{m^2} \left[\frac{1}{2} \sum_{v \in V} \sum_{i=2,5} (f(\sigma_i(v)) - f(v))^2 + \frac{1}{4} \sum_{v \in V} \sum_{i=1,3,4,6} (f(\sigma_i(v)) - f(v))^2 \right]$$

$$\leq \frac{1}{2m^2} \sum_{(u,v) \in E} (f(u) - f(v))^2.$$

Also $\int_T \phi^2 = \frac{1}{m^2} \sum_{v \in V} f^2(v)$. Therefore, by (6.4) we have

$$\frac{1}{2m^2} \sum_{(v,u) \in E} (f(v) - f(u))^2 \geq \frac{c}{m^2} \sum_{v \in V} f^2(v).$$

Thus,

$$\sum f = 0 \quad \text{implies} \quad \sum_{(v,u) \in E} (f(v) - f(u))^2 \geq 2c \langle f, f \rangle.$$

The claim is proved.   □

We can construct graphs with larger degrees and bounded eigenvalues by taking products of M_n as follows. The graph M_n^k has vertex set V, and two vertices u

and v are joined by s parallel edges where s is the number of walks of length k in M_n from v to u. Thus the adjacency matrix of M_n^k has eigenvalues ρ_i^k where ρ_i are eigenvalues of the adjacency matrix of M_n. Although this construction does not give as good eigenvalues as the following Ramanujan graphs, the construction schemes are simple and the approach is interesting.

Construction 6.3.6. The Ramanujan graphs $X^{p,q}$.

One of the major developments in constructive methods was the construction of Ramanujan graphs by Lubotzky, Phillips and Sarnak [**194**] and independently by Margulis [**196, 197, 198**]. Ramanujan graphs are k-regular graphs with λ_1 satisfying

$$\lambda_1 \geq 1 - \frac{2\sqrt{k-1}}{k}.$$

This eigenvalue bound is the best possible for fixed k where the number n of vertices approaches infinity (see Section 2.4).

The construction can be given for large classes of parameters for k and n using quotients of quaternion groups: : Let p be a prime congruent to 1 modulo 4 and let $H(Z)$ denote the integral quaternions

$$H(Z) = \{\alpha = a_o + a_1 i + a_2 j + a_3 k : a_j \in Z\}.$$

Let $\bar{\alpha} = a_0 - a_1 i - a_2 j - a_3 k$ and $N(\alpha) = \alpha\bar{\alpha} = a_0^2 + a_1^2 + a_2^2 + a_3^2$. It can be shown that there are precisely $\frac{p+1}{2}$ conjugate pairs $\{\alpha, \bar{\alpha}\}$ of elements of $H(Z)$ satisfying $N(\alpha) = p, \alpha \equiv 1 (\text{mod } 2)$ and $a_0 > 1$. Denote by S the set of all such elements. For each α in S, we associate the 2×2 matrix $\tilde{\alpha}$

$$\tilde{\alpha} = \begin{pmatrix} a_0 + ia_1 & a_2 + ia_3 \\ -a_2 + ia_3 & a_0 - ia_1 \end{pmatrix}$$

Let q be another prime congruent to 1 modulo 4. By taking the i in $\tilde{\alpha}$ to satisfy $i^2 \equiv -1 \pmod{q}$, $\tilde{\alpha}$ can be viewed as an element in $PGL(2, Z/qZ)$, which is the group of all 2×2 matrices over Z/qZ. Now we form the Cayley graph of $PGL(2, Z/qZ)$ relative to the above $p + 1$ elements. (The Cayley graph of a group G relative to a symmetric set of elements S is the graph with vertex set G and edges $\{x, y\}$ if $x = sy$ for some s in S.) If the Legendre symbol $\left(\frac{p}{q}\right) = 1$, then this graph is not connected since the generators all lie in the index two subgroup $PSL(2, Z/qZ)$, each element of which has determinant a square. So there are two cases. The Ramanujan graph $X^{p,q}$ is defined to be the above Cayley graph if $\left(\frac{p}{q}\right) = -1$, and to be the Cayley graph of $PSL(2, Z/qZ)$ relative to S if $\left(\frac{p}{q}\right) = 1$. For $\left(\frac{p}{q}\right) = -1$, $X^{p,q}$ is bipartite with edges between $PSL(2, Z/qZ)$ and its set theoretic complement. The Ramanujan graphs of interest here correspond to taking $\left(\frac{p}{q}\right) = 1$ and are $(p + 1)$-regular graphs with $q(q^2 - 1)/2$ vertices. The bipartite graphs with $\left(\frac{p}{q}\right) = -1$ are bipartite expander graphs.

The eigenvalue λ_1 can be determined to be $1 - \frac{2\sqrt{p}}{p+1}$ by using the results of Eichler [**109**] and Igusa [**164**] on the Ramanujan conjecture [**194, 217**] for the above case. In addition to having the optimum eigenvalue λ_1, Ramanujan graphs have many other nice properties. They serve as illuminating examples for various extremal problems, several of which we will discuss in Section 6.5.

6.4. Applications of expanders in communication networks

There are many applications of expander graphs scattered about in recent research papers in theoretical computer science. Expanders are extensively used in randomized and derandomized algorithms, computational complexity, and parallel architectures. There are many "tricks of the trade" in using expanders for amplifying a random bit, generating pseudorandom numbers, and constructing error-correcting codes, for example. We do not intend to cover these new applications since there are many more still being developed. Instead, we will focus on a classical example of using the power of expanders in communication networks.

Among various applications of expander graphs, those in communication networks have the longest history, and can be traced back to the early development of switching networks [**63, 195, 211, 212**]. Roughly speaking, a telephone network which provides connections for users is made of many parts or "gadgets", each of which performs some desired function. We will describe some of the gadgets that can be built by using expanders.

A *non-blocking network* is a directed graph with two specified disjoint subsets of vertices, one of which consists of input vertices and the other of output vertices. Now suppose that a number of calls take place simultaneously in the network, i.e., there are vertex-disjoint paths joining some inputs to outputs in the graph. Suppose one additional call comes in and it is desired to establish a new path joining the given input to the desired output without disturbing the existing calls, i.e., the new path is to be vertex-disjoint from the existing paths. The problem of interest is to minimize the number of edges in such a non-blocking network.

To build a non-blocking network, we need several types of building blocks, one of which is called a *k-access graph*, which has the property that for any given set S of vertex-disjoint paths connecting inputs to outputs, a new input can be connected to k different outputs by paths not containing any vertex in S. If k is greater than or equal to half of the total number of outputs, the k-access graph is a so-called *major access* network. A non-blocking network can then be formed by combining a major access network and its mirror image as shown in Fig. 1.

We construct here a major-access network $M(n)$ with n inputs and $24n$ outputs by combining 2 copies of $M(n/2)$ and 2 copies of the bipartite Ramanujan graphs $R(12n, 5)$ with $12n$ inputs and with degree $p + 1 = 6$, as illustrated in Fig. 2.

To verify that the above construction is a major-access network, we consider an input v which must have access to $6n$ of the middle vertices. After deleting the n possible vertices in S, the remaining set has at least $5n$ inputs of $M(n)$. In each of the Ramanujan graphs with degree $k = 6$ and $\lambda = 1 - \sqrt{5}/3$, we see that, by Lemma 6.4, the neighborhood of a subset S of size $5n$ has size at least:

$$\frac{5n}{(1 - \lambda)^2 + \frac{5}{12}\lambda(2 - \lambda)} = \frac{27n}{4}.$$

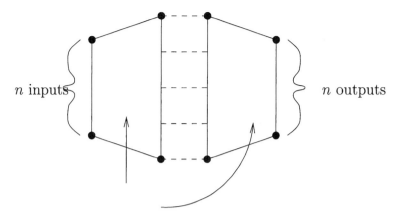

major-access network

FIGURE 1. A nonblocking network

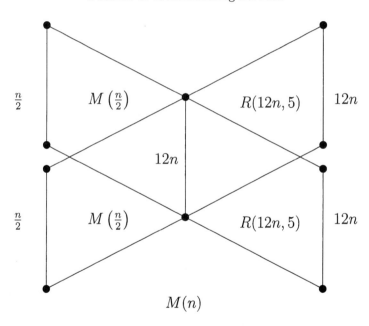

$M(n)$

FIGURE 2. A major access network

Since there are two copies of the Ramanujan graph $R(12n, 5)$ here, there are at least $\frac{27n}{2}$ neighbors of S. Among the $\frac{27}{2}n$ such outputs, there are at least $\frac{25}{2}n$ of them not in S, which is more than half of the outputs of $M(n)$. Therefore the above construction yields $M(n)$ and the number of edges of $M(n)$ satisfies

$$e(M(n)) = 2e(M(\frac{n}{2})) + 6 \cdot 12 \cdot 2n.$$

It can then be easily checked that the above major-access network has at most $144n \log n$ edges and therefore the nonblocking network has at most $288n \log n$ edges.

Another useful network is the so-called *superconcentrator*. Despite this impressive name, it actually has a very simple property. Namely, it is a graph with n inputs and n outputs, having the property that for any set of inputs and any set of outputs, a set of vertex-disjoint paths exists that join the inputs in a one-to-one fashion to the outputs (so that here it does not matter who is connected to whom.) The question of interest is to determine how few edges a superconcentrator can have. In fact, this has been taken as a measure for comparing the effectiveness of the expanders which are used to build superconcentrators. A simple recursive construction [**195**] of a superconcentrator is shown in Figure 3.

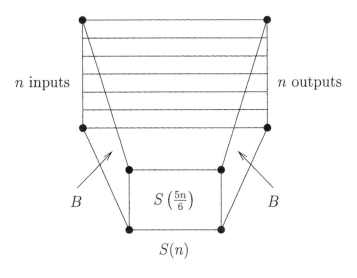

FIGURE 3. A superconcentrator

In the network in Figure 3, there is a matching between the n inputs and n outputs. Furthermore, the graph B has n inputs and $5n/6$ outputs satisfying the property that for any given $n/2$ inputs there is a set of vertex-disjoint paths joining the inputs in a one-to-one fashion to different outputs. For example, as defined in Section 6.2, an $(n, 5/6, k, 1/2, 1/2)$-concentrator has the above property. So for any given set of m inputs and m outputs in $S(n)$ of Figure 3, we can use the matching to provide $m - n/2$ disjoint paths and let the rest be achieved recursively in $S(\frac{5n}{6})$. Therefore the key part of the construction is made from an expander as in Figure 4.

In Figure 4, the first $n/6$ inputs, each having degree 5, are joined to $5n/6$ distinct outputs. The remaining $5n/6$ inputs are joined to the outputs by a Ramanujan graph with degree $6 = p + 1$. Now suppose we have a set of inputs X. It suffices to show that X has at least $\mid X \mid$ neighbors as outputs. Here we verify the situation for $\mid X \mid = n/2$ (the other cases where $\mid X \mid < n/2$ are easier). If X contains at least $\frac{n}{10}$ inputs among the first $\frac{n}{6}$ inputs, then we are done. We may assume X contains at least $\frac{2n}{5}$ inputs as an input set X' of the expander C. Since the expander graph has eigenvalue $\lambda = 1 - \sqrt{5}/3$, it is straightforward to check that

$$N(X') \geq \frac{\mid X' \mid}{\lambda(2 - \lambda)\frac{\mid X' \mid}{\frac{5}{6}n} + (1 - \lambda)^2} \geq \frac{9 \cdot \frac{2}{5}n}{4 \cdot \frac{12}{25} + 5} > n/2.$$

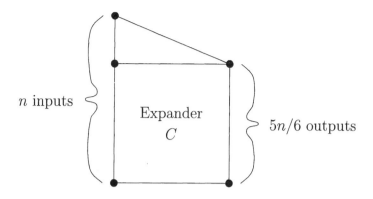

n inputs

Expander
C

$5n/6$ outputs

FIGURE 4. A concentrator

Now the total number of edges in the superconcentrator $S(n)$ satisfies

$$e(S(n)) \;=\; n + 2e(B) + e(S(5n/6))$$

$$=\; n + 2 \cdot 5/6n \cdot 7 + e(S(5n/6))$$

It is easy to verify that the above superconcentrator $S(n)$ has at most $76n$ edges. The number of edges in $S(n)$ can be reduced to $69.8n$ by replacing $S(\frac{5}{6}n)$ by $S((\frac{4}{5} + \epsilon)n)$ where $\epsilon = .0288776$, and in B each of the first $(\frac{1}{5} - \epsilon)n$ inputs of B has degree 4 or 5, and they are joined to a total of $(\frac{4}{5} + \epsilon)n$ distinct outputs of B.

It is worth mentioning that by using expanders, the existence of which is guaranteed by probabilistic methods, one can construct superconcentrators with as few as $36n$ edges. The parameters of such probabilistic expanders are given in Theorem 6.6 (also, see [**17**]). We remark that he best currently known lower bound for a superconcentrator of size n is $5n + O(logn)$, due to Lev and Valiant [**185**].

THEOREM 6.6. *For real numbers $0 < \alpha < 1/\beta < 1$, suppose that an integer k satisfies*

$$k > \frac{H(\alpha) + H(\alpha\beta)}{H(\alpha) - \alpha\beta H(1/\beta)}$$

where $H(x) = -x \log_2 x - (1 - x) \log_x (1 - x)$ is the entropy function. Then for any integer n, there exists a k-regular bipartite graph with vertex set $A \cup B$, $|A| = |B| = n$, so that every subset $X \subseteq A$ with $|X| \leq \alpha n$ has at least $\beta|X|$ neighbors in B.

In fact almost all random k-regular bipartite graphs between two sets of n vertices satisfy the above property. The proof is by combinatorial probabilistic methods and can be found in [**63**].

6.5. Constructions of graphs with small diameter and girth

There are many related extremal properties that are satisfied by random graphs but are somewhat "weaker" than the properties mentioned in Section 6.2. One such example is the diameter, which is the maximum distance between pairs of vertices.

There are graphs with small diameter but which do not have good expansion, discrepancy or eigenvalue properties.

The following two related extremal problems often arise in interconnection networks [110]:

Problem 6.5.1. Given k and D, construct a graph with as many vertices as possible with degree k and diameter D.

Problem 6.5.2. Given n and k, construct a graph with minimum degree on n vertices with diameter as small as possible.

It is not difficult to see that a graph with degree k and diameter D can have at most $M(k, D) = 1 + k + \cdots + k(k-1)^{i-1} + \cdots + k(k-1)^{D-1}$ vertices, which is sometimes called the Moore bound. Let $n(k, D)$ denote the maximum number of vertices in a graph with degree at most k and diameter at most D. Clearly, $n(k, D) \leq M(k, D)$. It has been shown that there are at most five kinds of graphs achieving the Moore bound, namely, cliques, odd cycles, the Petersen graph, the Hoffman-Singleton graph (with $k = 7$ and $D = 2$) and possibly the case of $M(2, 57)$ (which remains still unresolved [113, 161]).

A random graph has small diameter. To be specific, Bollobas and de la Vega [36] proved that a random k-regular graph has diameter $\log_{k-1} n + \log_{k-1} \log\, n + c$ for some small constant $c < 10$. This is almost the best possible in the sense that any k-regular graph has diameter at least $\log_{k-1} n$. Upper bounds for the diameter in terms of eigenvalues were discussed in Chapter 3. Namely, a k-regular graph G on n vertices has diameter at most $\lceil \frac{\log(n-1)}{\log(1/(1-\lambda))} \rceil$.

Using the above bound, the Ramanujan graph has diameter at most $\lceil \frac{\log n}{\log(1/(1-\lambda))} \rceil$ which comes to within a factor of 2 of the optimum diameter. In other words, the Ramanujan graph has the number of vertices about a factor of $(k-1)^{-D/2}$ times the Moore bound [161]. Quite a few other constructions, such as de Bruijn graphs [42] and their variations, fall within the range of 2^{-D} times the Moore bound. It remains an open problem to determine the maximum number $n(k, D)$ of vertices in a graph with degree k and diameter D. Relatively little is known about the upper bound for $n(k, D)$. The following somewhat trivial sounding question concerning the upper bound is still unresolved [113]:

Problem 6.5.3. Is it true that for every integer c, there exist k and D such that $n(k, D) < M(k, D) - c$?

There are many papers on this problem; the reader is referred to [21, 22, 46, 65, 66] for surveys on this topic.

Another direction is to allow additional edges to minimize diameter:

Problem 6.5.4. How small can the diameter be made by adding a matching to an n-cycle?

It was shown in [32] that by adding a random matching to an n-cycle the resulting graph has best possible diameter in the range of $\log_2 n$. In fact, a more general theorem can be proved which ensures that by adding a random matching to k-regular graphs, say Ramanujan graphs, the resulting graphs have diameter about $\log_{k-1} n$. It would be of interest to answer Problem 6.5.4 and its generalizations by explicit constructions [6, 18, 68, 133, 248].

Vijayan and Murty [248] asked the following problems related to reliability of graphs with diameter constraints:

Problem 6.5.5. Given n, D, D' and s, what is the minimum number of edges in a graph on n vertices of diameter D with the property that after removing s edges the remaining graph has diameter no more than D'?

These problems have attracted the attention of many researchers. There is a large literature on this problem [21, 22, 63, 65, 106] However, the answers are far from satisfactory and explicit solutions are few. Most problem of this type remain unsolved.

The *girth* of a graph is the size of a smallest cycle in the graph [26, 165, 253]. The girth of a random k-regular graph is known to be $\log_{k-1} n$ [117]. In [194], the bipartite Ramanujan graphs on n vertices have girth about $\frac{4}{3} \log_{k-1} n$; which is better than that of a random graph in the sense of avoiding small cycles. (To be precise, the bipartite Ramanujan graph $X^{p,q}$ has $q(q^2 - 1)$ vertices and girth $4 \log_p q - \log_p 4$.) This is closely related to the following extremal problem which, though rather old, remains open [37, 230]:

Problem 6.5.6. For a given integer $t \geq 6$, how many edges can a graph on n vertices have without containing any cycle of length $2t$?

Erdős conjectures that the maximum number $f(n, t)$ of edges in a graph on n vertices avoiding C_{2t} is of order $n^{1+1/t}$. It is not hard to see that $f(n, t) < n^{1+1/t}$. The Ramanujan graphs yield $f(n, t) > n^{1+2/3t}$ which is a substantial improvement upon previous lower bounds of $n^{1+\frac{1}{2t-1}}$ in [37]. Recently, Lazebnik, Ustimenko and Woldar [180] constructed graphs on n vertices, for infinitely many n, which have $n^{1+2/(3t-3)}$ edges containing no cycle of length $2t$.

6.6. Weighted Laplacians and the Lovász ϑ function

For a simple graph G and a weight matrix $W = (w_{uv})$ where $w_{uv} = w_{vu} \geq 0$ and $w_{uv} = 0$ for u and v not adjacent, we define the W-weighted Laplacian $\mathcal{L}_W$, a matrix with rows and columns indexed by vertices of G,

$$\mathcal{L}_W(u, v) = \begin{cases} 1 - \dfrac{w_{v,v}}{w_v} & \text{if } u = v, \text{ and } w_v \neq 0 \\[2mm] -\dfrac{w_{uv}}{\sqrt{w_u w_v}} & \text{if } u \sim v \text{ and } u \neq v, \\[2mm] 0 & \text{otherwise,} \end{cases}$$

where $w_v = \sum_u w_{uv}$.

Previously we have defined the Laplacian of a (fixed) weighted graph. The above definition is different in the sense that we start with a fixed graph G and consider all possible weight matrices.

For a graph G with a weight matrix W, let $\lambda_{max}(G_W)$ denote the maximum eigenvalue of the W-weighted Laplacian of G. In particular, let $\lambda_{max}(G)$ denote $\lambda_{max}(G_W)$ when the weight matrix satisfies $w_{uv} = 1$ if $u \sim v$ and 0 otherwise.

The σ function of a graph G is defined as follows:

$$\sigma(G) = 1 + \max_W \frac{1}{\lambda_{max}(G_W) - 1}$$

where W ranges over all weight matrices.

The chromatic number $\chi(G)$ of a graph G is the smallest integer k such that the vertices of G can be k-colored so that any two adjacent vertices have different colors. Bounds are known for $\chi(G)$ in terms of the eigenvalues of the adjacency matrix $A = (A_{uv})$ of G (where $A_{uv} = 1$ if u and v are adjacent, and 0 otherwise). Let $\rho_{\min}$ and $\rho_{\max}$ denote the smallest and largest eigenvalues of A, respectively. An upper bound of $\chi(G)$ was first given by H. S. Wilf [251] who showed that

$$\chi(G) \leq 1 + \rho_{\max}.$$

For the lower bound, A. J. Hoffman [162] proved (also see [154])

$$\chi(G) \geq 1 - \frac{\rho_{\max}}{\rho_{\min}} .$$

Here we will show that the σ function serves as a lower bound for the chromatic number $\chi(G)$ with a proof very similar to that in [162].

THEOREM 6.7. *For any graph G which contains at least one edge, we have*

$$\chi(G) \geq 1 + \max_W \frac{1}{\lambda_{max}(G_W) - 1} = \sigma(G)$$

PROOF. We consider the matrix M_1 with the (u, v)-entry to be $\sqrt{w_u w_v}$. If we regard $T^{1/2}\mathbf{1}$ as a row vector and $(T^{1/2}\mathbf{1})^*$ as a column vector, we can write

$$M_1 = (T^{1/2}\mathbf{1})(T^{1/2}\mathbf{1})^* = T^{1/2}JT^{1/2}.$$

where $w = \sum_v w_v$ and J is the all 1's matrix.

We consider eigenvalues of

$$M = \mathcal{L} - I + \frac{\lambda}{w}M_1$$

where $\lambda = \lambda_{max}(G_W)$ denotes the maximum eigenvalue of the W-weighted Laplacian $\mathcal{L}$ of G.

It is easy to see that the eigenfunctions of $\mathcal{L}$ are eigenfunctions of M and the maximum eigenvalue of M is $\lambda - 1$.

Let X denote an independent subset of G and let M_X denote the principal submatrix of M with rows and columns restricted to vertices in X. Then

$$M_X = \frac{\lambda}{w}(M_1)_X = \frac{\lambda}{w}(T^{1/2}\mathbf{1})_X(T^{1/2}\mathbf{1})_X^* \ .$$

The maximum eigenvalue of M is

$$\frac{\lambda}{w}(T^{1/2}\mathbf{1})_X^*(T^{1/2}\mathbf{1}) = \frac{\lambda}{w}w(X)$$

where $w(X) = \displaystyle\sum_{v \in X} w_v$.

Since the maximum eigenvalue of a principal submatrix of a symmetric matrix M is no more than the maximum eigenvalue of M, we have

$$\lambda - 1 \geq \frac{\lambda}{w}w(X) \ .$$

Suppose G has chromatic number $\chi(G) = k$. Thus, the vertex set of G can be partitioned into k independent sets $X_1, \ldots, X_k$. We then have

$$\lambda - 1 \geq \frac{\lambda}{w}w(X_i) \ , \ i = 1, \ldots, k \ .$$

In particular, the sum over all i satisfies

$$
\begin{aligned}
k(\lambda - 1) \ &\geq \ \frac{\lambda}{w}\sum_i w(X_i) \\
&= \ \lambda \ .
\end{aligned}
$$

This implies

$$\chi(G) = k \geq 1 + \frac{1}{\lambda - 1}$$

and (5.1) is proved. $\qquad\square$

As an immediate consequence of Theorem 6.7, we have the following corollaries:

COROLLARY 6.8. *For any graph G, we have*

$$\chi(G) \geq 1 + \max_{G'} \frac{1}{\lambda_{max}(G'_W) - 1}$$

where G' ranges over all induced subgraphs of G and W ranges over all weight matrices for G'.

We note that the inequality above is strict for some graphs (i.e., odd cycles for example). We also have the following simpler and weaker inequality.

COROLLARY 6.9. *For a graph G, we have*

$$\chi(G) \geq 1 + \frac{1}{\lambda_{max}(G) - 1}$$

where G' ranges over all induced subgraphs of G.

In a graph G, the *clique number* $\omega(G)$ of G is the number of vertices in G forms the largest complete subgraph (which is also called a clique). The problem of determining the clique number of a graph is known to be NP-complete [140]. A graph G is *perfect* if G and all its induced subgraphs have the property that the chromatic number equals the size of a largest clique. There is an excellent survey by Lovász [192] on perfect graphs. In general, the clique number and the chromatic number can be quite different. Here we will show that the σ function serves as an upper bound for the clique number (also see [224]).

THEOREM 6.10.

$$(6.5) \qquad\qquad \omega(G) \leq \sigma(G)$$

PROOF. Suppose the clique number of G is k. Clearly, $k > 1$. Let $S = \{v_1, \cdots, v_k\}$ denote the vertex set of the maximum clique. For each edge $\{u, v\}$, $u, v \in S$, we define

$$w_{uv} = w_{vu} = 1 .$$

Let $\theta \neq 1$ denote a k-th root of unity, say,

$$\theta = exp(\frac{2\pi i}{k}).$$

We consider $f(v) = \theta^j$ for v in C_j, $1 \leq j \leq k$. Then we have

$$\lambda_{max} \geq \frac{\displaystyle\sum_{u \sim v} |(f(u) - f(v))|^2 w_{uv}}{\displaystyle\sum_v |f(v)|^2 w_v}$$

$$= \frac{\displaystyle\sum_{j \neq m} |\theta^j - \theta^m|^2}{\displaystyle\sum_v w_v}$$

$$= \frac{\displaystyle\sum_j \sum_m (1 - (\theta^{-j+m} + \theta^{j-m})/2)}{k(k-1)}$$

$$= \frac{k^2}{k(k-1)}$$

$$= \frac{k}{k-1} .$$

Therefore

$$\omega(G) = k \leq 1 + \frac{1}{\lambda_{max} - 1} \leq \sigma(G)$$

and (6.5) is proved.   □

Combining (5.1) and (6.5), we have proved:

$$\omega(G) \leq \sigma(G) \leq \chi(G).$$

The σ function is closely related to Lovász's ϑ function which was first introduced in Lovász's seminal work on the Shannon capacity [191]. The ϑ function has many alternative definitions (see [151, 152]). One of the formulations states

$$\vartheta(G) = 1 + \max_{W'} \frac{\rho_{max}(W')}{\rho_{min}(W')}$$

where $\rho_{max}(W')$ denotes the maximum eigenvalue of W' and W' ranges over all matrices with rows and columns indexed by the vertex set of G satisfying $W'(u,v) \neq 0$ only if the vertices u and v are non-adjacent.

The original definition for $\vartheta(G)$ is as follows: For a graph G, an *orthonormal labeling* of G is an assignment of a unit vector x_v in a Euclidean space to each vertex v of G, in such a way that $\langle x_u, x_v \rangle = 0$ if $u \neq v$ and u is not adjacent to v. The ϑ function $\vartheta(G)$ is equal to:

$$\vartheta(G) = \min_{x,y} \max_{v} \frac{1}{\langle x_v, y \rangle^2}$$

where the minimum ranges over all orthonormal labelings x and all unit vectors y.

As pointed out by A. Galtman [138], the σ function can be expressed as :

(6.6) $$\sigma(G) = 1 + \max_{W} \frac{\rho_{max}(W)}{\rho_{min}(W)}$$

where W ranges over all matrices with rows and columns indexed by the vertex set of G satisfying $W(u,v) > 0$ only if the vertices u and v are adjacent.

Therefore, we have

$$\sigma(G) \leq \vartheta(\bar{G}).$$

The above formulation for $\sigma(G)$ coincides with that of a function introduced by McEliece, Rodemich and Rumsey [201] in their study of bounds for the independence number (In [201], the notation α_L was used). They proved that for a family of graphs the σ function is equal to the *Delsarte linear programming bound*. Independently, A. Schrijver [224] considered the σ function (and he used the notation ϑ'). The above inequality is strict for some graphs. In [224], an example by M.R. Best was given for a graph H satisfying $\sigma(H) \neq \vartheta(H)$. The vertex set of H is $\{0,1\}^6$ and two vertices are adjacent if and only if they differ in at most three of the six coordinates.

Schrijver gave the following definition for $\vartheta'(G) = \sigma(G)$: For a graph G, an *acute orthonormal labeling* of G is an assignment of a unit vector x_v in a Euclidean space to each vertex v of G, in such a way that $\langle x_u, x_v \rangle \geq 0$ if $u = v$ or u is adjacent to v and 0 otherwise. Then

$$\sigma(G) = \vartheta'(G) = \min_{x,y} \max_{v} \frac{1}{\langle x_v, y \rangle^2}$$

where the minimum ranges over all acute orthonormal labelings x and all unit vectors y.

So, altogether we have the so-called *Sandwich Theorem*:

$$\omega(G) \leq \sigma(G) \leq \vartheta(\bar{G}) \leq \chi(G).$$

How good are these inequalities? In other words, how effective can $\vartheta(G)$ or $\sigma(G)$ be used to estimate $\omega(G)$ or $\chi(G)$? Although the experimental evidence indicates that they are good in many cases, theoretical results show that significant obstacles are present in this approach.

Erdős [112] proved that

$$(6.7) \qquad \frac{\chi(G)}{\omega(G)} < \frac{cn}{(\log n)^2}$$

and for every n, there is a graph on n vertices satisfying

$$(6.8) \qquad \frac{\chi(G)}{\omega(G)} \geq \frac{c'n}{(\log n)^2}.$$

In fact, (6.8) is achieved by random graphs since Ramsey theory implies that a random graph has $\omega(G) \leq 2\log n/\log 2$ and therefore its chromatic number $\chi(G) \geq n/\omega(\bar{G}) \geq c'n/\log n$.

It has been shown ([124], also see [13, 122]) that for some $\epsilon > 0$, it is impossible to approximate in polynomial time the independence number of a graph on n vertices within a factor of n^ϵ, assuming $P \neq NP$. The exponent ϵ has recently been improved [158] , under similar assumptions, to $1 - \delta$ for every positive δ.

On the positive side, there have been a number of recent papers on approximation algorithms for coloring problems, max-cut problems and clique numbers using the ϑ function [11, 155, 141, 174]. Although these approximation algorithms often have complicated statements for certain ranges of the parameters with incremental or weak improvements, this approach still represents a significant step towards the understanding of these hard problems.

Notes:
The first four sections in this chapter are mainly based on a survey article [69]. Section 6.5 deals with several problems previously discussed in [65, 66, 68, 32]. Concerning the last section on the σ function and the chromatic number, the reader is referred to an excellent survey paper by Knuth [172]. Related material can also be found in [151, 152].

CHAPTER 7

Eigenvalues of symmetrical graphs

7.1. Symmetrical graphs

Roughly speaking, graphs with symmetry have "large" automorphism groups. There are various notions of symmetry in graphs such as vertex-transitivity, edge-transitivity, and distance transitivity (all of which we will soon define precisely). In general, the eigenvalues of symmetrical graphs have many remarkable properties, several of which we now mention:

1: For symmetrical graphs, eigenvalues can be easily bounded in terms of the diameter D and the degree k. For example, for vertex transitive graphs we have

$$(7.1) \qquad\qquad\qquad \lambda_1 \geq \frac{1}{kD^2}.$$

This implication has very interesting and important consequences. In many random walk problems, we might not even know the number of vertices in a graph, but the above inequality provides an efficient way to estimate the eigenvalues of a graph as long as the diameter and degree can be computed or bounded. This is a main component for the recent developments in rapidly mixing Markov chains [**228**]. Several eigenvalue inequalities with a form similar to (7.1) will be derived in a clean way in Section 7.2.

2: A graph is *distance transitive* if for any two pairs of vertices (x, y), (z, w) with $d(x, y) = d(z, w)$, there is a graph automorphism mapping x to z and y to w. Distance transitive graphs have very special spectra. We will describe a simple and natural method to determine the spectrum of a distance transitive graph. Basically, the spectrum of a distance transitive graph is the same as that of a corresponding contracted path with $D + 1$ vertices where D is the diameter. In fact, a distance transitive graph has exactly $D + 1$ distinct eigenvalues.

3: For some symmetrical graphs, eigenvalues can be explicitly determined by using group representations. We will discuss the general methods and give two examples: intersection graphs and the "Buckyball" graph (the truncated icosahedron). Intersection graphs have vertices labelled by subsets and adjacency determined by the cardinality of intersections of the sets corresponding to the endpoints. The Buckyball graph is associated with the newly found molecule "Buckminsterfullerene C_{60}", the third form of solid carbon. Not only does it have beautiful symmetry (which can be described by the alternating group A_5 or, equivalently, $PSL(2, 5)$), but it also provides a good example of an application of a graph theory model to chemistry.

Before we proceed, we start with some definitions. For a graph G, an *automorphism* $f : V(G) \to V(G)$ is a one-to-one mapping which preserves edges, i.e., for $u, v \in V(G)$, we have $\{u, v\} \in E$ if and only if $\{f(u), f(v)\} \in E$.

A graph G is *vertex-transitive* if its automorphism group $Aut(G)$ acts transitively on the vertex set $V(G)$, i.e. , for any two vertex u and v there is an automorphism $f \in Aut(G)$ such that $f(u) = v$. We sometimes write $f(u) = fu$ (especially when both f and u are represented by elements of a group).

We say that G is *edge-transitive* if, for any two edges $\{x, y\}, \{z, w\} \in E(G)$, there is an automorphism f such that $\{f(x), f(y)\} = \{z, w\}$. It is not difficult to show that an edge-transitive graph is either vertex-transitive or bipartite (or both).

A *homogeneous* graph is a graph Γ together with a group $\mathcal{H}$ which acts transitively on its vertex set. In other words, $\mathcal{H}$ is a subgroup of $Aut(\Gamma)$ and for any two vertices $u, v \in V(\Gamma)$ there is a $g \in \mathcal{H}$ such that $gu = v$. (Since G is the favorite common notation for both graph theorists and group theorists, we will try to avoid using G for either the graph or the group in connection with a homogeneous graph.) Any vertex-transitive graph Γ can be viewed as a homogeneous graph by taking $Aut(\Gamma)$ as the associated group. However, the choice of the associated group gives different ways of labelling the vertices in the following sense: In a homogeneous graph Γ with the associated group $\mathcal{H}$, the *isotropy* group $\mathcal{I}$ is defined as

$$\mathcal{I} = \{g \in \mathcal{H} : gv = v\}$$

for a fixed vertex v. We can identify V with the coset space $\mathcal{H}/\mathcal{I}$. The edge set of a homogeneous graph Γ can be described by an edge generating set $\mathcal{K} \subset \mathcal{H}$ so that each edge of Γ is of the form $\{v, gv\}$ for some $v \in V$ and $g \in \mathcal{K}$. For undirected graphs we require $g \in \mathcal{K}$ if and only if $g^{-1} \in \mathcal{K}$.

Cayley graphs are homogeneous graphs in which the isotropy group $\mathcal{I}$ is trivial, $\mathcal{I} = \{id\}$. In other words, the vertices of a Cayley graph are labelled by its associated group.

For example, the cycle C_n can be viewed as a homogeneous graph with associated group $\mathbb{Z}$ and isotropy group $n\mathbb{Z}$. Of course, C_n can also be viewed as a Cayley graph with associated group $\mathbb{Z}_n = \mathbb{Z}/n\mathbb{Z}$.

A symmetric graph is regular, so that its Laplacian satisfies

$$\mathcal{L} = I - \frac{1}{k} A$$

where k is the degree and A is the adjacency matrix.

7.2. Cheeger constants of symmetrical graphs

In this section, we will discuss several inequalities for Cheeger constants of vertex-transitive and edge-transitive graphs. The various versions of the following inequalities can be found in many places (see Babai and Szegedy [**14**]). Here we give a simple proof of a strengthened version.

THEOREM 7.1. *Suppose* Γ *is a finite edge-transitive graph of diameter* D. *Then the Cheeger constant* h_Γ *satisfies*

$$h_\Gamma \geq \frac{1}{2D}.$$

PROOF. Let S denote a subset of vertices such that $|S| \leq \frac{n}{2}$ where $n = |V(\Gamma)|$. We consider a random (ordered) pair of vertices (x, y), uniformly chosen over $V(\Gamma) \times V(\Gamma)$. Now, we choose randomly a shortest path P between x and y (uniformly chosen over all possible shortest paths). Since Γ is edge-transitive the probability that P goes through a given edge is at most $\frac{2D}{\text{vol } \Gamma}$.

A path between a vertex from S and a vertex from $\bar{S}$ must contain an edge in $E(S, \bar{S})$. Therefore we have

$$\frac{2|E(S, \bar{S})| \cdot D}{\text{vol } \Gamma} \geq Prob(x \in S, y \in \bar{S} \text{ or } x \in \bar{S}, y \in S).$$

This implies

$$\begin{aligned} |E(S, \bar{S})| &\geq \frac{|S|\,|\bar{S}|\text{vol } \Gamma}{Dn^2}. \\ \frac{E(S, \bar{S})}{k|S|} &\geq \frac{|\bar{S}|}{Dn} \\ &\geq \frac{1}{2D}. \end{aligned}$$

Therefore

$$h_\Gamma \geq \frac{1}{2D}.$$

$\square$

THEOREM 7.2. *Suppose* Γ *is a finite vertex-transitive graph of diameter* D *and degree* k. *Then the Cheeger constant* h_Γ *satisfies*

$$h_\Gamma \geq \frac{1}{2kD}.$$

PROOF. We follow the notation in the previous proof. The automorphism group defines an equivalence relation on the edges of Γ. Two edges e_1, e_2 are equivalent if and only if there is an automorphism π mapping e_1 to e_2. We can then consider equivalence classes of edges, denoted by $E_1, \cdots, E_s$.

We define the index of Γ to be

$$index\ \Gamma = \min_i \frac{\text{vol } \Gamma}{2|E_i|}$$

where E_i denotes the i-th equivalence class of edges. Clearly, we have $1 \leq index\ \Gamma \leq k$. In particular, when Γ is edge-transitive, we have $index\ \Gamma = 1$.

Since Γ is vertex transitive, each equivalence class contains at least $\frac{n}{2}$ edges. Let p_i denote the probability that a pair of vertices is an edge in the i-th equivalence class E_i. Since all edges in the same equivalence class have the same probability, we have, for each i,

$$p_i \leq \frac{1}{|E_i|} \leq \frac{2 \; index \; \Gamma}{\text{vol } \Gamma}.$$

For a subset S of the vertex set with $\text{vol} S \leq \text{vol} \bar{S}$, and, for a pair of vertices x, y in $\Gamma(G)$, the probability of having one of x, y in S and the other in $\bar{S}$ is the same as the probability that $P(x, y)$ contains an edge in $E(S, \bar{S})$. Therefore, we have

$$Prob(x \in S, y \in \bar{S} \; or \; x \in \bar{S}, y \in S) \;\; \leq \;\; |E(S, \bar{S})| D \; \max_i p_i$$

$$\leq \;\; |E(S, \bar{S})| D \cdot \frac{2 \; index \; \Gamma}{\text{vol } \Gamma}.$$

Since

$$Prob(x \in S, y \in \bar{S} \; or \; x \in \bar{S}, y \in S) = \frac{2|S| \; |\bar{S}|}{n^2}$$

we have

$$\frac{|E(S, \bar{S})|}{\text{vol } S} \geq \frac{1}{2D \; index \; \Gamma} \geq \frac{1}{2kD}$$

as claimed.   □

In fact, the above proof gives the following slightly stronger result.

THEOREM 7.3. *Suppose Γ is a finite vertex-transitive graph of diameter D. Then*

(7.2) $$h_\Gamma \geq \frac{1}{2D \; index \; \Gamma}.$$

We remark that in the inequality (7.2) the term *index* Γ cannot be deleted. There are, for example, Cayley graphs Γ with $h_\Gamma = \frac{c}{kD}$ for some constant c.

EXAMPLE 7.4. We consider the Cayley graph with vertex set $\mathbb{Z}_q \times \mathbb{Z}_2$ and edge generating set $\{(k, 0) : k \in \mathbb{Z}_q\} \cup \{(0, 1)\}$. In other words, the graph Γ is formed by joining two complete graphs on q vertices by a matching. This is the "dumb-bell" graph and its Cheeger constant satisfies

$$h = \frac{1}{q} = \frac{2}{kD} << \frac{1}{D}$$

where degree $k = q$ is large and diameter $D = 2$.

7.3. Eigenvalues of symmetrical graphs

We here derive lower bounds for eigenvalues of edge-transitive and vertex-transitive graphs. These bounds are somewhat stronger than the bounds one obtains for general graphs by using Cheeger inequalities.

THEOREM 7.5. *For an edge-transitive graph Γ with diameter D and degree k, we have*

$$\lambda_1 \geq \frac{1}{D^2}.$$

THEOREM 7.6. *For a vertex-transitive graph Γ with diameter D, we have*

$$\lambda_1 \geq \frac{1}{kD^2}.$$

The above two theorems are special cases of the following theorem:

THEOREM 7.7. *For a vertex transitive graph Γ with diameter D, we have*

$$\lambda_1 \geq \frac{1}{D^2 \; index \; \Gamma}$$

where index $\Gamma = \min \frac{vol \; \Gamma}{2|E_i|}$, and where E_i denotes the i-th equivalence class of edges under $Aut(\Gamma)$.

PROOF. We consider $f : V(\Gamma) \to \mathbb{R}$ and we use the (equivalent) eigenvalue definition (1.5) in Chapter 1:

$$\lambda_1 = \min_f \frac{n \sum\limits_{x \sim y} (f(x) - f(y))^2}{k \sum\limits_{x} \sum\limits_{y} (f(x) - f(y))^2}$$

For each edge $e = \{x, y\}$, we define

$$f(e) = |f(x) - f(y)|.$$

We then have

$$\lambda_1 = \min_f \frac{n \sum\limits_{e \in E} f^2(e)}{k \sum\limits_{x} \sum\limits_{y} (f(x) - f(y))^2}.$$

Let E_i denote the i-th equivalence class of edges under $Aut(\Gamma)$. For a fixed vertex x_0, we choose a fixed set of shortest paths $P_{x_0,y}$ to all y in Γ. We can now use the automorphism group to define, for each vertex $x \in V(\Gamma)$ and an automorphism π with $\pi(x_0) = x$, a set of paths

$$P(x) = \{\pi(P_{x_0,y})\}.$$

Clearly, each path in $P(x)$ has length at most D. For each edge e, we consider the number N_e of occurrences of e in paths in $P(x)$ ranging over all x.

Two edges in the same equivalence class have the same value for N_e. The total number of edges in all paths in $P(x)$ for all x is at most $n^2 D$. For each i and $e \in E_i$, we have

$$N_e \leq \frac{n^2 D}{2|E_i|} \leq \frac{n^2 D}{2 \min\limits_{i} |E_i|} = \frac{nD \; index \; \Gamma}{vol \; \Gamma} = \frac{D \; index \; \Gamma}{k}.$$

Now we consider, for a harmonic eigenfunction f achieving λ_1,

$$\sum_x \sum_y (f(x) - f(y))^2 \;=\; \sum_x \sum_y (\sum_{e \in P(x,y)} f(e))^2$$

$$\leq \sum_x \sum_y D \sum_{e \in P(x,y)} f^2(e)$$

$$\leq \sum_{e \in E} f^2(e) D N_e$$

$$\leq \sum_{e \in E} f^2(e) D \cdot \frac{D \; index \; \Gamma}{k}.$$

Therefore we have

$$\lambda_1 \;=\; \frac{n \sum f^2(e)}{k \sum_x \sum_y (f(x) - f(y))^2}$$

$$\geq \frac{1}{D^2 \; index \; \Gamma}.$$

This completes the proof of Theorem 7.7. $\square$

7.4. Distance transitive graphs

We recall that a graph Γ is distance transitive if for any two pairs of vertices $\{x, y\}$, $\{z, w\}$ with $d(x, y) = d(z, w)$, there is a graph automorphism f satisfying $f(x) = z$ and $f(y) = w$. In this section, all graphs we consider are distance transitive. We first point out an easy observation which has many useful consequences.

Observation: Let λ denote an eigenvalue of a distance transitive graph Γ. Suppose we fix a vertex v in Γ. Let $V_i = \{u \in \Gamma : d(u, v) = i\}$. Then λ is achieved by a nontrivial eigenfunction f with $f(v) \neq 0$ which satisfies that for each i, $x, y \in V_i$ implies $f(x) = f(y)$.

This observation follows from the fact that for any $x, y \in V_i$ there is an automorphism mapping x to y that fixes v. If we start from any eigenfunction f achieving λ, the new function formed by assigning the average value in V_i to vertices in V_i is also an eigenfunction with the same eigenvalue.

THEOREM 7.8. *The eigenvalues of a distance transitive graph Γ of diameter D are the eigenvalues of a weighted path P of $D + 1$ vertices.*

PROOF. We form a path P by contracting each set V_i as defined above into one vertex. The weight of the edge $\{v_i, v_{i+1}\}$ in P is just the number of edges between V_i and V_{i+1}. The loop $\{v_i, v_i\}$ has weight equal to twice the number of edges in Γ with both endpoints in V_i plus the number of loops in V_i. For an eigenfunction f of Γ with eigenvalue λ, we choose v so that $f(v)$ is nonzero and we map v into v_0 of the path P. Then we assign the average value of f (over the vertices in V_i) to v_i in P. Clearly, the resulting function is not always 0 and is a harmonic eigenfunction of P with the same eigenvalue λ. $\square$

It is, of course, much easier to deal with a $(D+1) \times (D+1)$ matrix of the path P with nonzero entries only at coordinates of $(i, i+1)$, (i, i), and $(i, i-1)$, than to deal with the Laplacian of G. An eigenfunction for P can be transformed into an eigenfunction for Γ as follows:

THEOREM 7.9. *A harmonic eigenfunction for the contracted path P is an eigenfunction for the distance transitive graph Γ.*

PROOF. Suppose f is a harmonic eigenfunction for P. We define $g(x) = f(v_i)$ if x is in V_i. It is easy to check that g is an eigenfunction for Γ. □

As a consequence of Theorem 7.8 and 7.9, we have

THEOREM 7.10. *A distance transitive graph of diameter D has exactly $D+1$ distinct eigenvalues.*

PROOF. From Theorem 7.9, we know that there are at most $D+1$ distinct eigenvalues. On the other hand, there are at least $D+1$ distinct eigenvalues because otherwise the characteristic polynomial $\prod\limits_{\lambda_i distinct} (\mathcal{L} - \lambda_i)$ would be of degree $\leq D$, contradicting the fact that $I, \mathcal{L}, \mathcal{L}^2, \dots, \mathcal{L}^d$ are independent. □

In fact, a slightly stronger statement is true.

THEOREM 7.11. *The eigenvalues of a distance transitive graph Γ of diameter D have the same values as the eigenvalues of a weighted path P of $D+1$ vertices, obtained by contracting each $V_i = \{u : d(u, v) = i\}$, for a fixed v, into a single vertex.*

Of course, the eigenvalues of Γ and P have different multiplicities. Nevertheless, there is a simple method to determine the multiplicities of eigenvalues of Γ from eigenfunctions of P. Note that in P, all eigenvalues have multiplicity 1.

THEOREM 7.12. *In a distance transitive graph Γ, we consider the $n \times n$ matrix A_j with $A_j(x, y) = 1$ if $d(x, y) = j$ in Γ and 0 otherwise. The eigenvalues for A_j are $|V_j| f(v_j) / f(v_0)$ for each harmonic eigenfunction f of P.*

PROOF. Since A_j is a polynomial in A, then an eigenfunction f of A is an eigenfunction for A_j. From Theorem 8.9, a harmonic eigenfunction f of P is an eigenfunction of A assuming the same value for vertices in V_j. Suppose the eigenvalue of f for A_j is λ'. We consider, for the vector $\varphi_0 = (1, 0, \dots, 0)$ of length $D+1$, the inner product

$$\langle \phi_0, \pi A_j f \rangle = \langle \phi_0, \lambda' \pi f \rangle = \lambda' f(v_0)$$

where π denotes the projection of $V(\Gamma)$ into P, i.e.,

$$\pi f(v_i) = \sum_{x \in V_i} f(x).$$

We remark that π can be viewed as a matrix with rows indexed by $V(P)$ and columns indexed by $V(\Gamma)$ such that $\pi(v, x) = 1$ if $x \in V_v$ and 0 otherwise. Clearly, we have

$$\langle \phi_0 \pi A_j, f \rangle = |V_j| f(v_j).$$

Therefore, $\lambda' = |V_j| f(v_j)/f(v_0)$. $\square$

Using Theorems 7.9, 7.10 and 7.12, we will show that the multiplicities of each eigenvalue can be determined from the eigenfunctions of the contracted path P.

THEOREM 7.13. *For a distance-transitive graph Γ on n vertices, the multiplicity $m(\lambda)$ of an eigenvalue λ is*

$$m(\lambda) = \frac{n}{\| g \|^2} g^2(v_0)$$

where g is the corresponding eigenfunction of the weighted path P contracted from Γ from a fixed vertex v_0.

PROOF. Let g_i's denote eigenfunctions of P and f_i's denote the corresponding harmonic eigenfunctions. Note that $g(v_i) = \sqrt{k|V_i|} f(v_i)$ where k is the degree of Γ. We consider the following $n \times n$ matrix:

$$M_i = \sum_{j=0}^{D} f_i(v_j) A_j.$$

For a fixed i, the trace of M_i is just the trace of $f_i(v_0) A_0$. Hence

$$(7.3) \qquad\qquad tr(M_i) = n f_i(v_0).$$

On the other hand, the eigenvalues of A_j are $|V_j| f_0(v_j)/f_0(v_0), |V_j| f_1(v_j)/f_1(v_0), \ldots,$ $|V_j| f_D(v_j)/f_D(v_0)$ of respective multiplicities $m(\lambda_0), \ldots, m(\lambda_D)$. Therefore,

$$
\begin{aligned}
tr\, M &= \sum_{j=0}^{D} f_i(v_j)\, tr\, A_j \\
&= \sum_{j=0}^{D} f_i(v_j) \sum_{l=0}^{D} m(\lambda_l) f_l(v_j) \frac{|V_j|}{f_l(v_0)} \\
&= \sum_{j=0}^{D} f_i(v_j) m(\lambda_i) f_i(v_j) \frac{|V_j|}{f_i(v_0)} \\
&= \| g_i \|^2 \frac{m(\lambda_i)}{f_i(v_0) k}.
\end{aligned}
$$

Here, we use the fact that distinct eigenfunctions are orthogonal so that

$$\sum_i f_p(v_i) f_q(v_i) |V_i| k = \begin{cases} 0 & \text{if } p \neq q \\ \| g_p \|^2 & \text{if } p = q. \end{cases}$$

Since $g^2(v_0) = k f^2(v_0)$, from (7.3), we have

$$m(\lambda) = \frac{n}{\| g \|^2} g^2(v_0).$$

The proof of Theorem 7.13 is complete. $\square$

EXAMPLE 7.14. *The Petersen graph and intersection graphs*

The Petersen graph is a distance transitive graph on 10 vertices. The vertex set is labelled by all 2-subsets of $\{1, 2, 3, 4, 5\}$. Two vertices A and B are adjacent if and only if $A \cap B = \emptyset$. The eigenvalues are 0, $\frac{2}{3}$ (with multiplicity 5), and $\frac{5}{3}$ (with multiplicity 4). The Petersen graph is a special case of the intersection graph $G(n, r, k)$ with vertex set consisting of all r-subsets of $\{1, \dots, n\}$. The vertices A and B are adjacent if $|A \cap B| = k$. The symmetric group S_n acts on the graph with isotropy group $S_r \times S_{n-r}$. Since $(S_n, S_r \times S_{n-r})$ is a Gelfand pair [97], the spectral decompositions of the space $\mathcal{F}(v) = \{f : V \to R\}$ are quite special:

$$\mathcal{F}(v) = E_0 \oplus E_1 \oplus \dots \oplus E_r$$

where the dimension of E_i satisfies $dim\, E_i = \binom{n}{i} - \binom{n}{i-1}$ for $i \geq 1$, and $dim\, E_0 = 1$.

7.5. Eigenvalues and group representation theory

A brute force method for computing eigenvalues of a connected graph on n vertices is to solve for x in the determinant, $det(xI - \mathcal{L})$, of an $n \times n$ matrix. Before starting such an arduous task, it makes sense to see if the matrix can be diagonalized into smaller blocks. Group representation theory is exactly the answer to such prayers when the graph is homogeneous.

Suppose Γ is a Cayley graph [25]with vertices labelled by a group $\mathcal{H}$ and with edge generating set $\mathcal{K}$. Let ρ denote an irreducible representation of $\mathcal{H}$ of dimension l. This means that ρ maps the elements of $\mathcal{H}$ into $l \times l$ matrices in such a way that matrix multiplication is consistent with the group multiplication, i.e., $\rho(g_1 g_2) = \rho(g_1)\rho(g_2)$. The eigenvalues of Γ are exactly the eigenvalues of the smaller matrix

$$I - \frac{1}{|k|} \sum_{g \in \mathcal{K}} \rho(g)$$

for ρ ranging over all irreducible representations of $\mathcal{H}$. Each eigenvalue of the $dim\, \rho \times dim\, \rho$ matrix has multiplicity $dim\, \rho$ in the graph Γ.

Suppose Γ is a homogeneous graph with associated group $\mathcal{H}$. The vertex set can be identified by $\mathcal{H}/\mathcal{I}$ where $\mathcal{I}$ is the isotropy group. The edge generating set $\mathcal{K} = \{g\mathcal{I} : v \sim gv\}$ for a fixed v is a union of double cosets

$$\mathcal{K} = \mathcal{I}\mathcal{K}\mathcal{I}$$

The eigenvalues of Γ are the eigenvalues of $I - \dfrac{1}{k} \sum_{g\mathcal{I} \in \mathcal{K}} \dfrac{1}{|\mathcal{I}|} \sum_{x \in g\mathcal{I}} \rho(x)$ where k is the degree of Γ.

The best way to illustrate the connection between homogeneous graph Γ and the irreducible representations of the associated group $\mathcal{H}$ is by examining concrete examples:

EXAMPLE 7.15. The cycle C_n as a Cayley graph associated with $\mathbb{Z}_n$. The Laplacian can be diagonalized since the irreducible representations $\rho_k(g) = \langle e^{\frac{2\pi i k}{n}}, g \rangle$, $k = 0, \dots, n-1$, are all 1-dimensional.

We remark that for the Gelfand pairs in Example 7.15, all irreducible representations are 1-dimensional. This simplifies the computation of the eigenvalues of the corresponding homogeneous graphs.

In our final example we use terminology that may be unfamiliar to some readers. A quick summary of this can be found in [85] or [84, 86].

EXAMPLE 7.16. The Buckyball, a soccer ball-like molecule, consists of 60 carbon atoms. It corresponds to a Cayley graph on A_5 with edge generating set $\{(12345), (54321), (12)(23)\}$. The edges generated by $(12)(34)$ correspond to "double bonds" and the edges generated by $(12345), (54321)$ to "single bonds". The irreducible representations for the alternating group were determined by Frobenius [134] and they are of dimensions $1, 3, 3, 4,$ and 5. This means the Laplacian can be diagonalized into blocks of sizes 1×1, 3×3 (with multiplicity 3), $3' \times 3'$ (a second type with multiplicity 3), 4×4 (with multiplicity 4), and 5×5 (with multiplicity 5). Note that $1^2 + 3^2 + 3^2 + 4^2 + 5^2 = 60$.

Suppose we consider the weighted graph with single bonds of weight 1 and double bonds of weight t. The eigenvalues of the adjacency matrix are exactly the eigenvalues of

$$\rho a + \rho a^{-1} + t \rho b$$

for any irreducible representation ρ and $a = (12345), b = (12)(34)$. For example, for the dimension 5 representation ρ_5 we have

$$\rho_5(a) = \begin{pmatrix} 0 & 1 & 0 & 0 & 0 \\ 0 & 0 & 1 & 0 & 0 \\ 0 & 0 & 0 & 1 & 0 \\ 0 & 0 & 0 & 0 & 1 \\ 1 & 0 & 0 & 0 & 0 \end{pmatrix}$$

$$\rho_5(a^{-1}) = \begin{pmatrix} 0 & 0 & 0 & 0 & 1 \\ 1 & 0 & 0 & 0 & 0 \\ 0 & 1 & 0 & 0 & 0 \\ 0 & 0 & 1 & 0 & 0 \\ 0 & 0 & 0 & 1 & 0 \end{pmatrix}$$

$$\rho_5(b) = \begin{pmatrix} -1 & -1 & -1 & -1 & -1 \\ 0 & 0 & 0 & 0 & 1 \\ 0 & 0 & 1 & 0 & 0 \\ 0 & 0 & 0 & 1 & 0 \\ 0 & 1 & 0 & 0 & 0 \end{pmatrix}$$

$$\rho_5(a) + \rho_5(a^{-1}) + t\rho_5(b) = \begin{pmatrix} -t & 1-t & -t & -t & 1-t \\ 1 & 0 & 1 & 0 & t \\ 0 & 1 & t & 1 & 0 \\ 0 & 0 & 1 & t & 1 \\ 1 & t & 0 & 1 & 0 \end{pmatrix}$$

Thus, the eigenvalues of the adjacency matrix are the roots of the characteristic polynomial

$$(x^2 + x - t^2 + t - 1)(x^3 - tx^2 - x^2 - t^2x + 2tx - 3x + t^3 - t^2 + t + 2)$$

.

In summary, the eigenvalues of the Buckyball can be written in closed form as roots of the following equations where the single bonds are weighted by 1 and the double bonds are weighted by t:

(a): $(x^2 + x - t^2 + t - 1)(x^3 - tx^2 - x^2 - t^2x + 2tx - 3x + t^3 - t^2 + t + 2) = 0$
with multiplicity 5;

(b): $(x^2 + x - t^2 - 1)(x^2 + x - (t + 1)^2) = 0$ with multiplicity 4;

(c): $(x^2 + (2t + 1)x + t^2 + t - 1)(x^4 - 3x^3 + (-2t^2 + t - 1)x^2 + (3t^2 - 4t + 8)x + t^4 - t^3 + t^2 + 4t - 4 = 0$ with multiplicity 3;

(d): $x - t - 2 = 0$ with multiplicity 1.

7.6. The vibrational spectrum of a graph

The Laplacian $\mathcal{L}$ of a graph Γ is an operator acting on the space of functions $\{f : V(\Gamma) \rightarrow \mathbb{R}\}$. A natural generalization is the *vibrational Laplacian* $\mathcal{L}_X$ which acts on the space $\mathcal{F}(V, X) = \{f : V(\Gamma) \rightarrow X\}$ for some vector space X. We use the word "vibrational" since the spectrum of the vibrational Laplacian $\mathcal{L}_X$ of a graph Γ for the special case of $X = \mathbb{R}^3$ is exactly the vibrational spectrum of the molecule whose atoms correspond to the vertices of Γ and whose bonds between atoms are just edges of Γ [84].

We start with a homogeneous graph Γ with the associated group $\mathcal{H}$ and isotropy group $\mathcal{I}$. We can generalize the Dirichlet sum (see Section 1.2) as follows: Suppose that to each edge, $e = \{u, v\}$, of the graph we associate a self-adjoint operator, A_e, on X; we then define the quadratic form Q on $\mathcal{F}(V, X)$ by

$$(7.4) \qquad \langle g, \mathcal{L}_X g \rangle = \frac{1}{2} \sum [g(u) - g(v)] \cdot A_e \cdot [g(u) - g(v)]$$

where the sum ranges over all edges of Γ. Suppose ρ denotes a representation of $\mathcal{H}$ on X. Furthermore, suppose A_e in (7.4) satisfies

$$A_{ae} = \rho(a) A_e \rho(a)^{-1}$$

where ae denotes the edge $\{ab, ac\}$ and the edge e is denoted by $\{b, c\}$. Then it can be shown (see [84]) that the spectrum of $\mathcal{L}_X$ can be decomposed into the union of the spectra of the following operators over all irreducible representations γ of Γ:

$$(7.5) \qquad \left(\sum_{g \in \mathcal{K}} A_g \right) \otimes I - \sum_{g \in \mathcal{K}} A_g \rho(g) \otimes \gamma(g)$$

where $\otimes$ denotes the cartesian product (of matrices).

Now we return to the example of the Buckyball graph. We will apply Hooke's law to derive the vibrational spectrum of this graph. Let $\mathbf{u} \in \mathbb{R}^3$ and $\mathbf{w} \in \mathbb{R}^3$ denote the equilibrium positions of the vertices labeled u and w. Let $h : V \rightarrow \mathbb{R}^3$

describe a deviation from equilibrium, so that $\mathbf{u} + h(u)$ is the new position of the vertex u. Then the potential energy associated to h can be expressed as:

$$W(h) = \frac{1}{2} \sum k_{u,w} (\|\mathbf{u} + h(u) - (\mathbf{w} + h(w))\| - \|\mathbf{u} - \mathbf{w}\|)^2.$$

In the above expression, the sum is over all pairs $\{u, w\}$ of vertices connected by an edge, and $k_{u,w}$ is the spring constant of that edge. If h is sufficiently small so as to enable us to ignore terms quadratic in h, we then have

$$\|\mathbf{u} + h(u) - (\mathbf{w} + h(w))\| \approx [\|\mathbf{u} - \mathbf{w}\|^2 + 2(\mathbf{u} - \mathbf{w}) \cdot (h(u) - h(w))]^{\frac{1}{2}}$$
$$\approx \|\mathbf{u} - \mathbf{w}\| + \omega_{u,w} \cdot (h(u) - h(w))$$

where

$$\omega_{u,w} = \frac{\mathbf{u} - \mathbf{w}}{\|\mathbf{u} - \mathbf{w}\|}$$

is the unit vector from $\mathbf{u}$ to $\mathbf{w}$ and $\cdot$ denotes the scalar product on $\mathbb{R}^3$. Then the quadratic approximation to W is given by

$$W(h) = \frac{1}{2} \sum k_{u,w} [\omega_{u,w} \cdot (h(u) - h(w))]^2.$$

Hence, we may take A_e to be a 3×3 matrix:

$$A_e = \omega_{u,w} \otimes \omega_{u,w}^t$$

where e is the edge joining u to w.

We can now use the above methods and (7.5) to compute explicitly the vibrational spectra of a molecule in terms of the irreducible representations of A_5. The space of displacements is $\mathcal{F}(V, \mathbb{R}^3) = \{f : V \to \mathbb{R}^3\}$. We choose ρ to be the ordinary three-dimensional representation (which is just rotation in $\mathbb{R}^3$). Using irreducible representations of A_5, we can then evaluate explicitly all vibrational eigenvalues by treating $3 \times 3, 9 \times 9, 9 \times 9, 12 \times 12$, and 15×15 matrices.

We point out that the above methods not only determine the vibrational spectrum, but also the specific representation associated to each eigenvalue. This additional information is important in chemical applications. For the case of homogeneous molecules, we can, in advance of all computations and independent of specific models for the potential energy, determine the number of representations of each type by a simple application of the Frobenius reciprocity formula.

Now the space of displacements of the Buckyball has dimension $180 = 60 \times 3$. But the space of entire (infinitesimal) rigid displacements of the molecule as a whole is six-dimensional (the Lie algebra of the Euclidean group). By subtracting these six dimensions, we get

1. *The space of vibrational states is 174-dimensional.*

The 180-dimensional space is the tensor product of the regular representation with a three-dimensional representation. So we must decompose the regular representation, which contains each irreducible representation with a multiplicity equal to its dimension. We have already mentioned the irreducibles of A_5 which we here denote as **1, 3, 3', 4** and **5**, where **3** is the three dimensional representation given by the action of A_5 on the icosahedron, and **3'** differs from **3** in that the generator of degree 5 has been replaced by its square. Since $1 + 3 + 3 + 4 + 5 = 16$, we see that the 180-dimensional displacement space decomposes into $48 = 3 \times 16$ irreducibles. Subtracting off two three-dimensional representations, we obtain:

2. *The number of distinct vibrational modes is at most 46.*

For a vibrational line to be visible as an absorption or emission line in the infrared (as a transition between the ground state and a one-photon state) it is necessary that the associated irreducible representation be equivalent to (the complexification of) the representation of $\mathcal{H}$ on the ordinary three-dimensional space $\mathbb{R}^3$ in which the molecule lies.

In the Raman experiment, light of a definite frequency is scattered with a change of frequency. This change, known as the Raman spectrum, is associated to those representations which intertwine with the space $S^2(\mathbb{R}^3)$ of symmetric two tensors. Therefore, both the infrared spectrum and Raman spectrum can be directly determined by using the Frobenius reciprocity formula. For details on this, the reader is referred to [84, 233].

3. *The space of classical vibrational states has dimension 174. Any force matrix, F, invariant under the group A_5 has (at most) 46 distinct eigenvalues yielding four lines visible in the infrared and ten in the Raman spectrum.*

Notes:

The proofs in Section 2 are mainly adapted from [14]. There are several chapters on distance transitive graphs in Biggs [25]. Here we have given slightly different proofs. The computation for the spectrum of the Buckyball graph is based on [87]. More reference on the vibrational spectrum of graphs can be found in [84, 85, 86].

CHAPTER 8

Eigenvalues of subgraphs with boundary conditions

8.1. Neumann eigenvalues and Dirichlet eigenvalues

In a graph G, for a subset S of the vertex set $V = V(G)$, the induced subgraph determined by S has edge set consisting of all edges of G with both endpoints in S. Although an induced subgraph can also be viewed as a graph in its own right, it is natural to consider an induced subgraph S as having a boundary. (Here and throughout, we shall denote by S the induced subgraph determined by S, when there is no danger of confusion.) There are two types of boundaries. The vertex boundary δS of an induced subgraph S consists of all vertices that are not in S but adjacent to some vertex in S (also defined in Chapter 2). The edge boundary, denoted by ∂S, consists of all edges containing one endpoint in S and the other endpoint not in S, but in the "host" graph. The host graph can be regarded as a special case of a graph with no vertex boundary and no edge boundary.

For an induced subgraph S with non-empty boundary, there are, in general, two kinds of eigenvalues — the Neumann eigenvalues and the Dirichlet eigenvalues, subject to different boundary conditions.

Neumann boundary condition:

The Laplacian $\mathcal{L}$ acts on functions $f : S \cup \delta S \to \mathbb{R}$ with the Neumann boundary condition, i.e., for every $x \in \delta S$

$$(8.1) \qquad \sum_{y \in S, y \sim x} (f(x) - f(y)) = 0.$$

Dirichlet boundary condition:

The Laplacian $\mathcal{L}$ acts on functions with the Dirichlet boundary condition. In other words, we consider the space of functions $\{f : S \cup \delta S \to \mathbb{R}\}$ which satisfy the Dirichlet condition

$$(8.2) \qquad f(x) = 0$$

for any vertex x in the vertex boundary δS of S.

We remark that the Dirichlet boundary condition for graphs corresponds naturally with that for smooth manifolds. In fact, the general condition with boundary value specified by a given function on δS will be discussed in the last section of this chapter. As we will see, problems with the general boundary condition can often be reduced to problems with the Dirichlet boundary condition (8.2).

The Neumann boundary condition (8.1) corresponds to the Neumann boundary condition for Riemannian manifolds:

$$\frac{\partial f(x)}{\partial \nu} = 0$$

for x on the boundary where ν is the normal direction orthogonal to the tangent hyperplane at x. Neumann eigenvalues are closely associated with random walk problems whereas the Dirichlet eigenvalues are related with many boundary-value problems.

8.2. The Neumann eigenvalues of a subgraph

Let S denote a subset of the vertex set $V(G)$ of G. Let S^* denote the union of the edges in S and the edges in ∂S. We define the Neumann eigenvalue of an induced subgraph S as follows:

$$(8.3) \qquad \lambda_S = \inf_{\substack{f \neq 0 \\ \sum_{x \in S} f(x)d_x = 0}} \frac{\displaystyle\sum_{\{x,y\} \in S^*} (f(x) - f(y))^2}{\displaystyle\sum_{x \in S} f^2(x)d_x}$$

$$= \inf_f \sup_c \frac{\displaystyle\sum_{\{x,y\} \in S^*} (f(x) - f(y))^2}{\displaystyle\sum_{x \in S} (f(x) - c)^2 d_x}$$

In general, we define the i-th Neumann eigenvalue $\lambda_{S,i}$ to be

$$\lambda_{S,i} = \inf_f \sup_{f' \in C_{i-1}} \frac{\displaystyle\sum_{\{x,y\} \in S^*} (f(x) - f(y))^2}{\displaystyle\sum_{x \in S} (f(x) - f'(x))^2 d_x}$$

where C_k is the subspace spanned by functions ϕ_j achieving $\lambda_{S,j}$, for $0 \leq j \leq k$. Clearly, $\lambda_{S,0} = 0$. We use the notation that $\lambda_{S,1} = \lambda_S$ and we remark that d_x still means the degree of x in G (independent of S).

From the discrete point of view, it is often useful to express the $\lambda_{S,i}$ as eigenvalues of a matrix $\mathcal{L}_S$. To achieve this, we first derive the following facts:

LEMMA 8.1. Let f denote a function $f : S \cup \delta S \to R$ satisfying (8.3) with eigenvalue λ. Then f satisfies:

(a): for $x \in S$,

$$Lf(x) = \sum_{\substack{y \\ \{x,y\} \in S^*}} (f(x) - f(y)) = \lambda f(x)d_x,$$

(b): *for $x \in \delta S$,*

$$\sum_{\substack{y \\ \{x,y\} \in \partial S}} (f(x) - f(y)) = 0.$$

This is the Neumann condition. Equivalently,

$$f(x) = \frac{1}{d'_x} \sum_{\substack{y \\ \{x,y\} \in \partial S}} f(y)$$

where d'_x denotes the number of neighbors of x in S.

(c): *for any function $h : S \cup \delta S \to \mathbb{R}$, we have*

$$\sum_{x \in S} h(x) L f(x) = \sum_{\{x,y\} \in S^*} (h(x) - h(y)) \cdot (f(x) - f(y)).$$

We remark that the proofs of (a) and (b) follow by variational principles and that (c) is a consequence of (b).

Using Lemma 8.1 and equation (8.3), we can rewrite (8.3) by considering the operator acting on the space of functions $\{f : S \to \mathbb{R}\}$, or, alternatively, on the space of functions $\{f : S \cup \delta S \to \mathbb{R}$ and f satisfies the Neumann condition$\}$:

$$(8.4) \qquad \lambda_S = \inf_{\substack{f \neq 0 \\ \sum f(x)d_x = 0}} \frac{\displaystyle\sum_{x \in S} f(x) L f(x)}{\displaystyle\sum_{x \in S} f^2(x) d_x}$$

$$= \inf_{g \perp T^{1/2} \mathbf{1}} \frac{\displaystyle\sum_{x \in S} g(x) \mathcal{L} g(x)}{\displaystyle\sum_{x \in S} g(x)^2}$$

$$= \inf_{g \perp T^{1/2} \mathbf{1}} \frac{\langle g, \, \mathcal{L} g \rangle_S}{\langle g, \, g \rangle_S}$$

where $\mathcal{L}$ is the Laplacian for the host graph G and $\langle f_1, f_2 \rangle_S = \displaystyle\sum_{x \in S} f_1(x) f_2(x)$.

For $X \subset V$, we let L_X denote the submatrix of L restricted to the columns and rows indexed by vertices in X. We define the following matrix N with rows indexed by vertices in $S \cup \delta S$ and columns indexed by vertices in S.

$$N(x, y) = \begin{cases} 1 & \text{if } x = y \\ 0 & \text{if } x \in S \text{ and } x \neq y \\ \dfrac{1}{d'_x} & \text{if } x \in \delta S, y \in S \text{ and } x \sim y \\ 0 & \text{otherwise} \end{cases}$$

Further, we define an $|S| \times |S|$ matrix

$$\mathcal{N}_S = T^{-1/2} N^* L_{S \cup \delta S} N T^{-1/2}$$

where N^* denotes the transpose of N.

It is easy to see from equation (8.4) that the $\lambda_{S,i}$ are exactly the Neumann eigenvalues of $\mathcal{N}_S$. In fact $\mathcal{N}_S$ has an eigenvalue 0 and, if S is connected, it has $|S| - 1$ positive eigenvalues.

8.3. Neumann eigenvalues and random walks

For an induced subgraph S of a graph G, we consider the following so-called Neumann random walk: The probability of moving from a vertex v in S to a neighbor u of v is $1/d_v$ if u is in S. If u is not in S, we then move from v to each neighbor of u in S with the (additional) probability $1/d_v d'_u$ where d'_u denotes the number of neighbors of u in S. The transition matrix P for this walk, whose columns and rows are indexed by S, is defined as follows:

$$(8.5) \qquad fP(v) = \sum_{\substack{u \in S \\ u \sim v}} \frac{1}{d_u} f(u) + \sum_{\substack{u \in S \\ u \sim z \sim v \\ z \notin S}} \frac{1}{d_u d'_z} f(u)$$

The stationary distribution is $d_v / \sum_u d_u$ at the vertex v. We remark that above Neumann walk is somewhat different from the random walks often used (in which if v has r neighbors not in S, then the probability of staying at v is r/d_v.) In a way, the Neumann walk takes advantage of "reflecting" from the boundary as dictated by the Neumann boundary condition.

The eigenvalues ρ_i of the transition probability matrix P associated with the Neumann walk are related to the Neumann eigenvalues $\lambda_{S,i}$ as follows:

$$\rho_i = 1 - \lambda_{S,i}.$$

In particular, we have

$$(8.6) \qquad \rho = \rho_1 = 1 - \lambda_{S,1} = 1 - \lambda_S.$$

This can be proved by using the Neumann condition as follows:

$$
1 - \rho \;=\; \inf_{f \neq 0} \frac{\displaystyle\sum_{\substack{x \sim y \\ x,y \in S}} (f(x) - f(y))^2 + \sum_{\substack{x \sim z \sim y \\ x,y \in S, z \notin S}} (f(x) - f(y))^2 / d_z}{\displaystyle\sum_{x \in S} f^2(x) d_x}
$$

$$
\geq \;\inf_{f \neq 0} \frac{\displaystyle\sum_{\substack{x \sim y \\ x,y \in S}} (f(x) - f(y))^2 + \sum_{z \notin S} \sum_{\substack{x \sim z \\ x \in S}} [d'_z f^2(x) - (\sum_{\substack{y \sim z \\ y \in S}} f(y))^2] / d'_z}{\displaystyle\sum_{x \in S} f^2(x) d_x}
$$

$$
\geq \;\inf_{f \neq 0} \frac{\displaystyle\sum_{\substack{x \sim y \\ x,y \in S}} (f(x) - f(y))^2 + \sum_{z \notin S} \sum_{\substack{x \sim z \\ x \in S}} (f^2(x) - f^2(z))}{\displaystyle\sum_{x \in S} f^2(x) d_x}
$$

$$
\geq \;\inf_{f \neq 0} \frac{\displaystyle\sum_{\substack{x \sim y \\ x,y \in S}} (f(x) - f(y))^2 + \sum_{z \notin S} \sum_{\substack{x \sim z \\ x \in S}} (f(x) - f(z))^2}{\displaystyle\sum_{x \in S} f^2(x) d_x}
$$

$$
= \;\inf_{f \neq 0} \frac{\displaystyle\sum_{\{x,y\} \in \hat{S}} (f(x) - f(y))^2}{\displaystyle\sum_{x \in S} f^2(x) d_x}
$$

$$
= \;\lambda_S
$$

where f ranges over all functions $f : \delta S \cup S \to \mathbb{R}$ satisfying

$$
\sum_{x \in S} f(x) d_x = 0,
$$

and for $x \in \delta S$

$$
\sum_{y \in S, y \sim x} (f(x) - f(y)) = 0.
$$

Inequality (8.6) is quite useful in bounding the rate of convergence of Markov chains for problems which can be formulated as random walks in a subgraph with a boundary.

Suppose S is an induced subgraph of a k-regular graph. The above random walk can be described as follows: At an interior vertex v of S, the probability of moving to each neighbor is equal to $1/k$. (An interior vertex of S is a vertex adjacent to no vertex outside of S.) At a boundary vertex $v \in \delta S$, the probability of moving to a neighbor u of v is $1/k$ if u is in S, and, in this case where $u \notin S$, an (additional) probability of $1/(kd'_u)$ is assigned for moving from v to each neighbor of u in S. The stationary distribution of the above random walk is just the uniform distribution.

8.4. Dirichlet eigenvalues

We consider a graph G with vertex set $V = V(G)$ and edge set $E = E(G)$. Let S denote a subset of V and we assume that the vertex boundary δS is nonempty.

We use the notation $f \in D^*$ to denote that f satisfies the Dirichlet boundary condition

$$f(x) = 0 \text{ for } x \in \delta S.$$

The Dirichlet eigenvalues of an induced subgraph on S are defined as follows:

$$(8.7) \qquad \lambda_1^{(D)} = \inf_{\substack{f \neq 0 \\ f \in D^*}} \frac{\displaystyle\sum_{\{x,y\} \in S^*} (f(x) - f(y))^2}{\displaystyle\sum_{x \in S} f^2(x) d_x}$$

$$= \inf_{\substack{g \neq 0 \\ g \in D^*}} \frac{\displaystyle\sum_{\{x,y\} \in S^*} \left(\frac{g(x)}{\sqrt{d_x}} - \frac{g(y)}{\sqrt{d_y}}\right)^2}{\displaystyle\sum_{x \in S} g(x)^2}$$

$$= \inf_{\substack{g \neq 0 \\ g \in D^*}} \frac{\langle g, \mathcal{L}g \rangle}{\langle g, g \rangle}$$

In general, we define the i-th Dirichlet eigenvalue λ_i to be

$$\lambda_i^{(D)} = \inf_{f \neq 0} \sup_{f' \in C_{i-1}} \frac{\displaystyle\sum_{\{x,y\} \in S^*} (f(x) - f(y))^2}{\displaystyle\sum_{x \in S} (f(x) - f'(y))^2 d_x}$$

where C_i is the subspace spanned by eigenfunctions ϕ_j achieving λ_j, for $1 \leq j \leq i$. We use the notation $\lambda_1^{(D)} = \lambda_S^{(D)}$. We will index the Dirichlet eigenvalues for a subgraph on $|S|$ vertices by $1 \leq i \leq |S|$.

With very similar proofs to those in Chapter 1, it can be shown that for a connected induced subgraph S of a graph G with $\partial S \neq \emptyset$, we have

$$0 < \lambda_1^{(D)} \leq \frac{|\partial S|}{\text{vol } S} \leq 1$$

$$0 < \lambda_i^{(D)} \leq 2$$

for $1 \leq i \leq |S|$.

LEMMA 8.2. *For an induced subgraph S, let g denote an eigenfunction of $\mathcal{L}$ with Dirichlet eigenvalue λ, i.e., $g : S \to \mathbb{R}$ satisfies (8.7) and the Dirichlet boundary condition $g(x) = 0$ for $x \in \delta S$. Then g satisfies*

(1): *for $x \in S$,*

$$\mathcal{L}g(x) = \frac{1}{\sqrt{d_x}} \sum_{\substack{y \\ \{x,y\} \in S^*}} \left(\frac{g(x)}{\sqrt{d_x}} - \frac{g(y)}{\sqrt{d_y}}\right) = \lambda g(x)$$

(2): *for any function $h : V \to \mathbb{R}$,*

$$\sum_{x \in S} h(x)\mathcal{L}g(x) = \sum_{\{x,y\} \in S^*} \left(\frac{h(x)}{\sqrt{d_x}} - \frac{h(y)}{\sqrt{d_y}}\right) \cdot \left(\frac{g(x)}{\sqrt{d_x}} - \frac{g(y)}{\sqrt{d_y}}\right)$$

The proof of (1) follows from the variational principles and will be omitted. To see (2), we note that

$$\sum_{x \in S} h(x)\mathcal{L}g(x) - \sum_{\{x,y\} \in S^*} \left(\frac{h(x)}{\sqrt{d_x}} - \frac{h(y)}{\sqrt{d_y}}\right) \cdot \left(\frac{g(x)}{\sqrt{d_x}} - \frac{g(y)}{\sqrt{d_y}}\right)$$

$$= \sum_{x \in \delta S} \frac{g(x)}{\sqrt{d_x}} \left(\frac{h(x)}{\sqrt{d_x}} - \frac{h(y)}{\sqrt{d_y}}\right)$$

$$= 0.$$

As an immediate consequence of Lemma 8.2, for functions $f : S \to \mathbb{R}$ under the assumption that $f(y) = 0$ for $y \in \delta S$, we have, for all $x \in S$,

$$\mathcal{L}f(x) = \mathcal{L}_S f(x)$$

where $\mathcal{L}_S$ is the submatrix of $\mathcal{L}$ restricted to columns and rows indexed by vertices in S. We note that since $\delta S \neq \emptyset$, $\mathcal{L}_S$ is nonsingular. All eigenvalues of $\mathcal{L}_S$ are positive. Hence The Dirichlet eigenvalues of the induced subgraph on S are just the eigenvalues of $\mathcal{L}_S$ and the determinant of $\mathcal{L}_S$ can be expressed as:

$$\det \mathcal{L}_S = \prod_{i=1}^{|S|} \lambda_i^{(D)}.$$

8.5. A matrix-tree theorem and Dirichlet eigenvalues

One of the classical combinatorial theorems is the matrix-tree theorem which states that the determinant of any cofactor of the combinatorial Laplacian is equal to the number of spanning trees in a graph. This result can be traced back to Kirchhoff [173]; see also Maxwell [199], Sylvester [235] and Cayley [47]. There is a large literature of related work such as Tutte [242], Bott and Mayberry [39], Trent [241], Uhlenbeck and Ford [243], Rényi [221], and Moon [203].

In this section, we consider a generalization of the matrix-tree theorem for induced subgraphs of a graph (see [89]). For an induced subgraph S with non-empty boundary in a graph G, we define a rooted spanning forest of S to be any subgraph F satisfying:
(1) F is an acyclic subgraph of G,
(2) F has vertex set $S \cup \delta S$,
(3) Each connected component of F contains exactly one vertex in δS.

The following theorem (see [**89**]) relates the product of the Dirichlet eigenvalues of S to the enumeration of rooted spanning forests of S:

THEOREM 8.3. *For an induced subgraph S in a graph G with $\delta S \neq \emptyset$, the number of rooted spanning forests of S is*

$$\prod_{x \in S} d_x \prod_{i=1}^{|S|} \lambda_i$$

where λ_i, $1 \leq i \leq |S|$, are the Dirichlet eigenvalues of the Laplacian of S in G.

PROOF. We consider the incidence matrix B with rows indexed by vertices in S and columns indexed by edges in S^* defined as follows:

$$B(x,e) = \begin{cases} \dfrac{1}{\sqrt{d_x}} & \text{if } e = \{x,y\}, x < y \\ -\dfrac{1}{\sqrt{d_x}} & \text{if } e = \{x,y\}, x > y \\ 0 & \text{otherwise.} \end{cases}$$

We have

(8.8) $$\mathcal{L} = B\ B^*$$

where B^* denotes the transpose of B.

Then

$$\begin{aligned} \prod_{i=1}^{|S|} \lambda_i &= \det \mathcal{L} \\ &= \det B\ B^* \\ &= \sum_X \det B_X \det B_X^* \end{aligned}$$

where X ranges over all possible choices of $s - 1$ edges and B_X denotes the square submatrix of B whose $s - 1$ columns correspond to the edges in X.

Claim 1: If the subgraph with vertex set $S \cup \delta S$ and edge set X contains a cycle, then $\det B_X = 0$.
The proof follows from the fact that the columns restricted to those indexed by the cycle are dependent.

Claim 2: If the subgraph formed by edge set X contains a connected component having two vertices in δS, then $\det B_X = 0$.
Proof: Let Y denote a connected component of the subgraph formed by X. If Y contains more than one vertex in δS, then Y has no more than $|E(Y)| - 1$ vertices in S. The submatrix formed by the columns corresponding to edges in Y has rank at most $|E(Y)| - 1$. Consequently, $\det B_X = 0$.

Claim 3: If the subgraph formed by X is a rooted forest of S, then

$$|\det B_X| = \frac{1}{\prod_{x \in S} \sqrt{d_x}}.$$

Proof: From Claims 1 and 2, we know that edges of X form a forest and each connected component contains exactly one vertex in δS. There is a column indexed by an edge with only one nonzero entry, say (x_1, e_1) with $x_1 \in S$. Therefore,

$$|\det B_x| = \frac{1}{\sqrt{d x_1}} |\det B_{x_1}^{(1)}|$$

where $B_{x_1}^{(1)}$ denotes the submatrix with rows indexed by $S - \{x_1\}$ and columns indexed by $X - \{e\}$. By removing one edge and one vertex at a time, we eventually obtain

$$|\det B_x| = \frac{1}{\prod_{x \in S}} \sqrt{d_x}.$$

Combining Claims 1-3, we have

$$\prod_{i=1}^{|S|} \lambda_i \;=\; \det \mathcal{L} = \sum_X \det B_X \det B_X^*$$

$$= \frac{1}{\prod_{x \in S} d_x} |\{\text{rooted spanning forests of } S\}|.$$

This completes the proof of Theorem 8.3.   □

We remark that the usual matrix-tree theorem can be viewed as a special case of Theorem 8.3. Namely, for a graph G, we apply Theorem 8.3 to an induced subgraph H on $V(G) - \{v\}$ for some vertex v in G. The rooted spanning forests correspond in an one-to-one fashion to all trees in G.

8.6. Determinants and invariant field theory

Conformal invariant theory for free bosons intimately uses the determinant of the Laplacian with Dirichlet boundary conditions (see [142], [178]). The discrete version can be described as follows: For an induced subgraph S with non-empty boundary δS, suppose σ is a function defined on the boundary δS. (Often the function σ satisfies some periodic conditions as well.) The "energy" for a function f is related to

$$H(f) = \sum_{\substack{x \sim y \\ x \in S}} [f(x) - f(y)]^2$$

where the summation ranges over all edges $\{x, y\}$ at least one endpoint of which is in S. The partition function can be expressed as

$$Z(\sigma) = \int e^{-c\, H(f)}$$

where f ranges over all functions whose restriction on δS is σ.

To compute $Z(\sigma)$, we note that the function f_0 that minimizes $H(f)$ and shares the same value as σ on the boundary, also satisfies the following condition for all $x \in S$:

$$\sum_{\substack{y \\ y \sim x}} (f_0(x) - f_0(y)) = 0.$$

This can be proved by variational principles. Also, it is not difficult to show that such a function exists if S is connected and, when it exists, is uniquely determined. For any function g whose restriction to δS is σ, we consider

$$f = g - f_0$$

Clearly, f satisfies the Dirichlet condition. We can rewrite $H(g)$ as follows:

$$
\begin{aligned}
H(g) &= \sum_{\{x,y\} \in S^*} (f(x) - f(y))^2 + \sum_{\{x,y\} \in S^*} (f_0(x) - f_0(y))^2 \\
&\quad + \sum_{\{x,y\} \in S^*} (f(x) - f(y))(f_0(x) - f_0(y)) \\
&= \sum_{\{x,y\} \in S^*} (f(x) - f(y))^2 + \sum_{\{x,y\} \in S^*} (f_0(x) - f_0(y))^2 \\
&\quad + \sum_{x \in S^*} f_0(x) \sum_{\substack{y \in S \\ x \sim y}} (f(x) - f(y)) \\
&= \sum_{\{x,y\} \in S^*} (f(x) - f(y))^2 + \sum_{\{x,y\} \in S^*} (f_0(x) - f_0(y))^2
\end{aligned}
$$

Therefore, we have

$$Z = e^{-c\, H(f_0)} \int_f e^{-c\, H(f)}$$

where f ranges over all functions satisfying the Dirichlet condition. If we express f relative to the basis formed by orthonormal eigenfunctions ϕ_i of the induced graph S with n vertices, we obtain

$$f = \sum_i x_i \phi_i.$$

Thus,

$$H(f) = \sum_i x_i^2 \lambda_i$$

and

$$
\begin{aligned}
Z &= e^{-c\, H(f_0)} \int e^{-c \sum_i \lambda_i x_i^2} dx_i \\
&= c^{-n/2} \Big(\prod_{i=1}^{|S|} \lambda_i \Big)^{-1/2} \pi^{-n/2}.
\end{aligned}
$$

Therefore the present problem is reduced to the problem of evaluating the product $\prod_{i=1}^{|S|} \lambda_i$ of the Dirichlet eigenvalues which, in turn, is the same as enumerating rooted spanning forests of an induced subgraph.

The problem of evaluating the determinants of Laplacians on Riemann surfaces has been considered by many researchers (e.g., Polyakov [215] in string theory, Ray-Singer [219] in analytic torsion, D'Hoker and Phong [95] in connection with Selberg's zeta function, Sarnak [223], Osgood, Phillips and Sarnak [206, 207] for isospectral sets). The discrete parallel involves a combinatorial trace formula and zeta functions, which however will not be included in this book because of space limitations. The heat kernel of a graph plays a major role there and will be discussed in the next chapter.

Notes:

The Neumann eigenvalues and Neumann walks are mainly based on [**60**] and [**79**]. More discussion on Dirichlet eigenvalues can be found in [**178**] and also in [**89**].

CHAPTER 9

Harnack inequalities

9.1. Eigenfunctions

A crucial part of spectral graph theory concerns understanding the behavior of eigenfunctions. Intuitively, an eigenfunction maps the vertices of a graph to the real line in such a way that edges serve as "elastic bands" with the effect of pulling adjacent vertices closer together. To be specific, let f denote a harmonic eigenfunction with eigenvalue λ in a graph G (or for an induced subgraph S with nonempty boundary). Locally, at each vertex, the eigenfunction stretches the incident edges in a balanced way. That is, for each vertex x, f satisfies

$$\sum_{\substack{y \\ y \sim x}} (f(x) - f(y)) = \lambda f(x) d_x.$$

Globally, we would like to have some notion that adjacent vertices are close to one another. In spectral geometry, Harnack inequalities are exactly the tools for capturing the essence of eigenfunctions. There are many different versions of Harnack inequalities (involving constants depending on the dimension of the manifold, for example.) We consider the following inequality for graphs:

Harnack inequality:
For every vertex x in a graph G and some absolute constant c, any eigenfunction f with eigenvalue λ satisfies

$$(9.1) \qquad \frac{1}{d_x} \sum_{\substack{y \\ y \sim x}} (f(x) - f(y))^2 \le c\lambda \max_z f^2(z).$$

Nevertheless, the above inequality in general does not hold for all graphs. An easy counterexample is the graph formed by joining two complete graphs of the same size by a single edge.

The main goal of this chapter is to establish the Harnack inequality (9.1) for certain homogeneous graphs and their subgraphs. In the next section, we will discuss several notions of convexity for graphs. After we establish a Harnack inequality for certain homogeneous graphs, we will examine Harnack inequalities for Neumann eigenvalues and Dirichlet eigenvalues for various convex subgraphs. We will also use the Harnack inequality to derive lower bounds for eigenvalues in terms of the diameter and degree of the graph. This section is mainly based on [**58**] and [**56**].

9.2. Convex subgraphs of homogeneous graphs

Let Γ denote a homogeneous graph with an associated group $\mathcal{H}$ and isotropy group $\mathcal{I}$. Let $\mathcal{K}$ denote the edge generating set; we require $g \in \mathcal{K}$ if and only if $g^{-1} \in \mathcal{K}$. We say that a homogeneous graph is *invariant* if for every element $a \in \mathcal{K}$, we have

$$a \mathcal{K} a^{-1} = \mathcal{K}.$$

In other words, $\mathcal{K}$ is invariant as a set under conjugation by elements of $\mathcal{K}$. For example, a homogeneous graph associated with an abelian group is invariant and we call the graph an abelian homogeneous graph. Suppose a homogeneous graph Γ is associated with a nonabelian group $\mathcal{H}$. If the edge generating set $\mathcal{K}$ is a subgroup of $\mathcal{H}$, then Γ is still invariant.

Before we give definitions for convex subgraphs, some explanations are in order. Although the following various notions of convexity are quite natural and interesting by themselves, they are dictated by the proofs given later for establishing the Harnack inequalities for Neumann eigenvalues and for Dirichlet eigenvalues. Another motivation in defining convex subgraphs is in its relevance with random walk problems. Many combinatorial enumeration and sampling problems involve families of combinatorial objects which can be regarded as vertices of convex subgraphs of appropriate host graphs. Harnack inequalities and the resulting eigenvalue bounds can be used to derive polynomial time upper bounds for the rate of convergence for the sampling problem, which often then give efficient approximation algorithms for the associated enumeration problems.

Strongly convex subgraphs

An induced subgraph S of a homogeneous graph Γ with vertex boundary δS is said to be *strongly convex* if the edge generating set $\mathcal{K}$ satisfies the following condition:
(A): For all $a, b \in \mathcal{K}$, $x \in \delta S$, if $ax \in S$, $bx \in S$, then we have $b^{-1}a \in \mathcal{K}$.

We note that (A) is implied by the following condition:

(B): For all pairs of vertices u and v in S, all shortest paths joining u and v are contained in S.

Clearly, the intersection of two strongly convex subgraphs is strongly convex.

EXAMPLE 9.1. Suppose we consider a homogeneous graph Γ_t with vertex set $\{(a_1, a_2, \cdots, a_t) \ : \ a_i \in \mathbb{Z}, \sum_i a_i = n\}$. A vertex $(a_1, \cdots, a_i, \cdots, a_j, \cdots, a_t)$ is adjacent to $(a_1, \cdots, a_i + 1, \cdots, a_j - 1, \cdots, a_t)$ for $1 \leq i, j \leq t$. In particular, for a fixed i, we define $H_i = \{(a_1, a_2, \cdots, a_t) \ : \ a_i \geq 0\}$ to be a halfplane. It is easy to see that H_i is strongly convex and $\cap H_i = \{(a_1, a_2, \cdots, a_t) \ \in \ V(\Gamma_t) : a_i \geq 0\}$ is strongly convex.

Convex subgraphs

An induced subgraph S of a homogeneous graph Γ with vertex boundary δS is said to be *convex* if for any subset $X \subset \delta S$, we have

$$(9.2) \qquad |N^*(X) \setminus (S \cup \delta S)| \geq |X|$$

where $N^*(X) = \{y \; : \; y \in X \text{ or } y \sim x \in X\}$. In other words, any subset X of the boundary δS of S has at least as many neighbors outside of $S \cup \delta S$ as the cardinality of X. We will call (9.2) the boundary expansion property.

LEMMA 9.2. *If two induced subgraphs F_1, F_2 are both convex, then the induced subgraph $F_1 \cap F_2$ is convex.*

PROOF. Suppose $X \subset \delta(F_1 \cap F_2)$. We can partition X into two parts $X_1 = X \cap \delta F_1$ and $X_2 = X \setminus \delta F_1$. Clearly X_2 is contained in F_1. Since F_1 and F_2 are convex, we have

$$|N^*(X_1) \setminus (F_1 \cup \delta F_1)| \geq |X_1|$$

$$|N^*(X_2) \setminus (F_2 \cup \delta F_2)| \geq |X_2|.$$

Since $N^*(X_2) \setminus F_2 \subset F_1 \cup \delta F_1$, we have

$$(N^*(X_2) \setminus (F_2 \cup \delta F_2)) \cap (N^*(X_1) \setminus (F_1 \cup \delta F_1)) = \emptyset.$$

Hence

$$
\begin{aligned}
|N^*(X) \setminus (F_1 \cap F_2 \cup \delta(F_1 \cap F_2))| &\geq |N^*(X_1) \setminus (F_1 \cup \delta F_1)| + |N^*(X_2) \setminus (F_2 \cup \delta F_2)| \\
&\geq |X_1| + |X_2| = |X|.
\end{aligned}
$$

and the proof is complete. □

EXAMPLE 9.3. We consider the space S of all $m \times n$ matrices with non-negative integral entries having column sums $c_1, \ldots, c_n$, and row sums $r_1, \ldots r_m$.
First, we construct a homogeneous graph Γ with vertex set consisting of all $m \times n$ matrices with integral (possibly negative) entries. Two vertices u and v are adjacent if they differ at four entries in some submatrix determined by two columns i, j and rows k, m satisfying

$$u_{ik} = v_{ik} + 1, u_{jk} = v_{jk} - 1, u_{im} = v_{im} - 1, u_{jm} = v_{jm} + 1$$

It is easy to see that Γ is a homogeneous graph with an edge generating set consisting of all 2×2 submatrices $\begin{pmatrix} 1 & -1 \\ -1 & 1 \end{pmatrix}$. A halfplane consists of matrices the (i, j)th entry of which is non-negative for some i and j. It is easy to see that S is just the intersection of halfplanes.

We remark that the definitions given here regarding convex subgraphs are far from being exhaustive. One simple and general notion of convexity is to embed the vertices of the host graph into a manifold. A convex subgraph is just an induced subgraph on the intersection of the vertex set of the host graph with some submanifold having convex boundary. The techniques in this section are not enough to derive Harnack inequalities for such a general definition of convex subgraph. We will return to this topic after we discuss the heat kernels of graphs in the next chapter.

9.3. A Harnack inequality for homogeneous graphs

We will first prove the following Harnack inequality for homogeneous graphs which are invariant.

THEOREM 9.4. *In an invariant homogeneous graph Γ with edge generating set $\mathcal{K}$ consisting of k generators, suppose a function $f : V(\Gamma) \to \mathbb{R}$ satisfies*

$$Lf(x) = \frac{1}{k} \sum_{a \in \mathcal{K}} [f(x) - f(ax)] = \lambda f(x).$$

Then the following inequality holds for all $x \in V(\Gamma)$ and $\alpha > 2$:

$$\frac{1}{k} \sum_{a \in \mathcal{K}} [f(x) - f(ax)]^2 + \alpha \lambda f^2(x) \leq \frac{\lambda \alpha^2}{\alpha - 2} \sup_y f^2(y).$$

PROOF. We define

$$\rho(x) = \frac{1}{k} \sum_{g \in \mathcal{K}} [f(x) - f(gx)]^2$$

and we consider

$$
\begin{aligned}
L\rho(x) &= \frac{1}{k^2} \sum_{b \in \mathcal{K}} \sum_{a \in \mathcal{K}} \{[f(x) - f(ax)]^2 - [f(bx) - f(abx)]^2\} \\
&= -\frac{1}{k^2} \sum_{b \in \mathcal{K}} \sum_{a \in \mathcal{K}} [f(x) - f(ax) - f(bx) + f(abx)]^2 \\
&\quad + \frac{2}{k^2} \sum_{b \in \mathcal{K}} \sum_{a \in \mathcal{K}} [f(x) - f(ax) - f(bx) + f(abx)][f(x) - f(ax)].
\end{aligned}
$$

Let X denote the last double summation above. Then

$$
\begin{aligned}
X &= \frac{2}{k^2} \sum_{b \in \mathcal{K}} \sum_{a \in \mathcal{K}} [f(x) - f(ax) - f(bx) + f(abx)] \, [f(x) - f(ax)] \\
&= \frac{2}{k^2} \sum_{a \in \mathcal{K}} \{\sum_{b \in \mathcal{K}} [f(x) - f(ax) - f(bx) + f(bax)]\}[f(x) - f(ax)] \\
&\quad + \frac{2}{k^2} \sum_{a \in \mathcal{K}} [\sum_{b \in \mathcal{K}} (f(abx) - f(bax))][f(x) - f(ax)] \\
&= \frac{2\lambda}{k} \sum_{a \in \mathcal{K}} [f(x) - f(ax)]^2 + \frac{2}{k^2} \sum_{a \in \mathcal{K}} [\sum_{b \in \mathcal{K}} (f(abx) - f(bax))][f(x) - f(ax)].
\end{aligned}
$$

Since Γ is an invariant homogeneous graph, we have

$$\sum_{b \in \mathcal{K}} (f(abx) - f(bax)) = 0$$

and therefore

$$L\rho(x) \leq X = \frac{2\lambda}{k} \sum_{a \in \mathcal{K}} [f(x) - f(ax)]^2.$$

Now we consider

$$
\begin{aligned}
Lf^2(x) &= \frac{1}{k}\sum_{a\in\mathcal{K}}[f^2(x) - f^2(ax)] \\
&= \frac{2}{k}\sum_{a\in\mathcal{K}} f(x)[f(x) - f(ax)] - \frac{1}{k}\sum_{a\in\mathcal{K}}[f(x) - f(ax)]^2 \\
&= 2\lambda f^2(x) - \frac{1}{k}\sum_{a\in\mathcal{K}}[f(x) - f(ax)]^2.
\end{aligned}
$$

Combining the above arguments, we have, for any positive α:

$$
L(\frac{1}{k}\sum_{a\in\mathcal{K}}[f(x) - f(ax)]^2 + \alpha\lambda f^2(x)) \le 2\alpha\lambda^2 f^2(x)
$$

$$
-\frac{(\alpha - 2)\lambda}{k}\sum_{a\in\mathcal{K}}[f(x) - f(ax)]^2.
$$

Now we consider a vertex v which maximizes, over all $x\in S$, the expression

$$
\frac{1}{k}\sum_{a\in\mathcal{K}}[f(x) - f(ax)]^2 + \alpha\lambda f^2(x).
$$

We have

$$
\begin{aligned}
0 &\le L(\frac{1}{k}\sum_{a\in\mathcal{K}}[f(v) - f(av)]^2 + \alpha\lambda f^2(v)) \\
&\le 2\alpha\lambda^2 f^2(v) - \frac{\lambda(\alpha - 2)}{k}\sum_{a\in\mathcal{K}}[f(v) - f(av)]^2.
\end{aligned}
$$

This implies

$$
\frac{1}{k}\sum_{a\in\mathcal{K}}[f(v) - f(av)]^2 \le \frac{2\lambda\alpha}{\alpha - 2}f^2(v)
$$

for $\alpha > 2$. Therefore for every $x\in V(\Gamma)$, we have

$$
\begin{aligned}
\frac{1}{k}\sum_{a\in\mathcal{K}}[f(x) - f(ax)]^2 + \alpha\lambda f^2(x) &\le \frac{1}{k}\sum_{a\in\mathcal{K}}[f(v) - f(av)]^2 + \alpha\lambda f^2(x) \\
&\le \frac{2\lambda\alpha}{\alpha - 2}f^2(v) + \alpha\lambda f^2(x) \\
&\le \frac{\alpha^2}{\alpha - 2}\lambda\max_y f^2(y)
\end{aligned}
$$

for $\alpha > 2$. The proof of Theorem 9.4 is complete. $\square$

By taking $\alpha = 4$ in Theorem 9.4 we have

THEOREM 9.5. *In an invariant homogeneous graph Γ with edge generating set $\mathcal{K}$ consisting of k generators, suppose a function $f : V(\Gamma) \to \mathbb{R}$ satisfies*

$$
Lf(x) = \frac{1}{k}\sum_{a\in\mathcal{K}}[f(x) - f(ax)] = \lambda f(x).
$$

Then for all $x \in V(\Gamma)$,

$$\frac{1}{k} \sum_{a \in \mathcal{K}} [f(x) - f(ax)]^2 \leq 8\lambda \sup_y f^2(y).$$

9.4. Harnack inequalities for Dirichlet eigenvalues

In this section, we consider Laplacians acting on functions with Dirichlet boundary conditions. Let S denote a convex subgraph of a graph G. We first prove a useful lemma which which is a consequence of convexity.

LEMMA 9.6. *For a convex subgraph S of a graph G, an eigenfunction f (with Dirichlet eigenvalue λ) defined on $S \cup \delta S$, can be extended to all vertices of G which are adjacent to some vertex x in $S \cup \delta S$ so that it satisfies*

$$\sum_{\substack{y \\ y \sim x}} (f(x) - f(y)) = \lambda f(x) d_x.$$

PROOF. First we note that any eigenfunction f must have value 0 for any $x \in \delta S$. To extend f to all vertices adjacent to some vertex in $S \cup \delta S$, we consider a system of $|\delta S|$ equations :

$$\sum_{\substack{y \\ y \sim x}} (f(x) - f(y)) = \lambda f(x) d_x$$

for each $x \in \delta S$. The variables are $f(z)$ for every $z \notin S \cup \delta S$ and $z \sim y \in S \cup \delta S$. The boundary expansion condition (9.2) implies that any k equations involve at least k variables for $k \leq |\delta S|$ and thus assures that this system of equations has solutions. $\qquad \square$

THEOREM 9.7. *Suppose S is a convex subgraph in an abelian homogeneous graph Γ with edge generating set $\mathcal{K}$ consisting of k generators. Let $f : S \to \mathbb{R}$ denote an eigenfunction with associated Dirichlet eigenvalue λ. Then the following inequality holds for $x \in S$ and $a \in \mathcal{K}$:*

$$[f(x) - f(ax)]^2 + k\alpha\lambda f^2(x) \leq \frac{k\lambda\alpha^2}{\alpha - 2} \sup_{y \in S} f^2(y)$$

for any $\alpha > 2$.

PROOF. Using Lemma 9.6, we can extend f to all vertices adjacent to some vertex in $S \cup \delta S$. For $g \in \mathcal{K}$, we define

$$\phi_g(x) = [f(x) - f(gx)]^2 + k\alpha\lambda f^2(x)$$

for $x \in S \cup \delta S$. First we note that the maximum value of $\phi_g(x)$, ranging over all $g \in \mathcal{K}$, $x \in S \cup \delta S$, is achieved by some edge $\{z, az\}$ where $z \in S$.

Let $\phi(x)$ denote $\phi_a(x)$. We consider, for $x \in S$,

$$
\begin{aligned}
L\phi(x) &= \frac{1}{k}\sum_{b \in \mathcal{K}}[\phi(x) - \phi(bx)]\\
&\leq \frac{1}{k}\sum_{b \in \mathcal{K}}[(f(x) - f(ax))^2 - (f(bx) - f(abx))^2] + \alpha\lambda\sum_{b \in \mathcal{K}}[f^2(x) - f^2(bx)]\\
&= Y + Z.
\end{aligned}
$$

We examine

$$
\begin{aligned}
Y &= \frac{1}{k}\sum_{b \in \mathcal{K}}[(f(x) - f(ax))^2 - (f(bx) - f(abx))^2]\\
&= \frac{2}{k}\sum_{b \in \mathcal{K}}(f(x) - f(ax) - f(bx) + f(abx))(f(x) - f(ax)\\
&\quad -\frac{1}{k}\sum_{b \in \mathcal{K}}(f(x) - f(ax) - f(bx) + f(abx))^2\\
&\leq \frac{2}{k}\{\sum_{b \in \mathcal{K}}[f(x) - f(bx)] - \sum_{b \in \mathcal{K}}[f(ax) - f(abx)]\}(f(x) - f(ax))\\
&\leq 2\lambda[f(x) - f(ax)]^2
\end{aligned}
$$

since $Lf(x) = \lambda f(x)$ for $x \in S \cup \delta S$. Also, we have

$$
\begin{aligned}
Z &= \alpha\lambda\sum_{b \in \mathcal{K}}[f^2(x) - f^2(bx)]\\
&\leq \alpha\lambda\{\sum_{b \in \mathcal{K}}2[f(x) - f(bx)]f(x) - \sum_{b \in \mathcal{K}}[f(x) - f(bx)]^2\}\\
&\leq \alpha\lambda\{2k\lambda f^2(x) - \sum_{b \in \mathcal{K}}[f(x) - f(bx)]^2\}.
\end{aligned}
$$

In particular, for v achieving the maximum value of ϕ, we have

$$
0 \leq L\phi(z) \leq 2k\alpha\lambda^2 f^2(z) - \lambda(\alpha - 2)\sum_{a \in \mathcal{K}}[f(z) - f(az)]^2
$$

and

$$
[f(z) - f(az)]^2 \leq \frac{2k\lambda\alpha}{\alpha - 2}f^2(z)
$$

for $\alpha > 2$. Therefore for all $x \in S, g \in \mathcal{K}, gx \in S$, we have

$$
\begin{aligned}
[f(x) - f(gx)]^2 + k\alpha\lambda f^2(x) &\leq [f(z) - f(az)]^2 + k\alpha\lambda f^2(x)\\
&\leq \frac{2k\lambda\alpha}{\alpha - 2}f^2(z) + k\alpha\lambda f^2(x)\\
&\leq \frac{k\lambda\alpha^2}{\alpha - 2}\max_{y \in S}f^2(y)
\end{aligned}
$$

for any $\alpha > 2$. The proof of Theorem 9.7 is complete.   □

By taking $\alpha = 4$ in Theorem 9.7 we have

THEOREM 9.8. *Suppose S is a convex subgraph in an abelian homogeneous graph Γ with edge generating set $\mathcal{K}$ consisting of k generators. Let $f : S \to \mathbb{R}$ denote an eigenfunction with associated Dirichlet eigenvalue λ. Then for all $x \in S, a \in \mathcal{K}$, we have*

$$[f(x) - f(ax)]^2 \le 8k\lambda \sup_{y \in S} f^2(y).$$

9.5. Harnack inequalities for Neumann eigenvalues

In this section, we focus on Laplacians acting on functions which satisfy the Neumann boundary condition. For strongly convex subgraphs, a similar but slightly different Harnack inequality for Neumann eigenvalues can be derived. The proof is based on the assumption that the homogeneous graph is abelian and uses the Neumann boundary condition.

THEOREM 9.9. *Suppose S is a finite strongly convex subgraph in an abelian homogeneous graph Γ with edge generating set $\mathcal{K}$ consisting of k generators. Let $f : S \to \mathbb{R}$ denote an eigenfunction with associated Neumann eigenvalue λ. Then the following inequality holds for $x \in S$, $a \in \mathcal{K}$ and $ax \in S$:*

$$[f(x) - f(ax)]^2 + k\alpha\lambda f^2(x) \le \frac{k\lambda\alpha^2}{\alpha - 2} \sup_{y \in S} f^2(y)$$

for any $\alpha > 2$.

PROOF. We consider, for some $g \in \mathcal{K}$,

$$\phi_g(x) = [f(x) - f(gx)]^2 + k\alpha\lambda f^2(x).$$

Let $\{z, az\}$ be an edge in S at which the maximum value of $\phi_g(x)$ ranging over all $g \in \mathcal{K}$, $x, gx \in S$, is achieved.

Claim: $\phi_g(x) \le \phi_a(z)$ for all $x, gx \in \delta S \cup S$.

Proof: We consider the following three possibilities:

Case 1: Suppose $gx \in S$ and $x \in \delta S$. The Neumann condition implies that

$$f(x) = \frac{1}{w} \sum_{b \in W} f(bx)$$

where $W = \delta x \cap S$ and $w = |W|$. Therefore we have

$$
\begin{aligned}
\phi_g(x) &= [f(x) - f(gx)]^2 + k\alpha\lambda f^2(x) \\
&= [\frac{1}{w} \sum_{b \in W} f(bx) - f(gx)]^2 + k\alpha\lambda[\frac{1}{w} \sum_{b \in W} f(bx)]^2.
\end{aligned}
$$

Thus

$$\phi_g(x) \le \frac{1}{w} \sum_{b \in W} [f(bx) - f(gx)]^2 + \frac{k\alpha\lambda}{w} \sum_{b \in W} f(bx)^2.$$

From the definition of strongly convex subgraph, bx is adjacent to gx for all $bx \in W$. Therefore, we have

$$\phi_g(x) \leq \frac{1}{w} \sum_{b \in W} \max_g \phi_g(bx) \leq \phi_a(z).$$

Case 2: Suppose $gx \in \delta S$ and $x \in S$.

$$
\begin{aligned}
\phi_g(x) &= [f(x) - f(gx)]^2 + k\alpha\lambda f^2(x) \\
&= [f(x) - \frac{1}{w'} \sum_{b \in W'} f(bgx)]^2 + k\alpha\lambda f^2(x) \\
&\leq \frac{1}{w'} \sum_{b \in W'} [f(x) - f(bgx)]^2 + k\alpha\lambda f^2(x) \\
&\leq \frac{1}{w'} \sum_{b \in W'} \max_g \phi_g(x) \\
&\leq \phi_a(z)
\end{aligned}
$$

where $W' = \delta gx \cap S$ and $w' = |W'|$. Here we use the fact that bgx is adjacent to x because of strong convexity.

Case 3: Suppose that $x, gx \in \delta S$. We define

$$
\begin{aligned}
W_1 &= \{b : bx \in S, bgx \in S\}, \\
W_2 &= \{b : bx \in S, bgx \notin S\}, \\
W_3 &= \{b : bx \notin S, bgx \in S\}.
\end{aligned}
$$

We set $w_i = |W_i|$ and $w = w_1 + w_2 + w_3$. Then

$$
\begin{aligned}
\phi_g(x) &= [f(x) - f(gx)]^2 + k\alpha\lambda f^2(x) \\
&= [\frac{w_3}{w} f(x) + \frac{1}{w} \sum_{b \in W_1 \cup W_2} f(bx) - \frac{w_2}{w} f(gx) - \frac{1}{w} \sum_{b \in W_1 \cup W_3} f(bgx)]^2 \\
&\quad + k\alpha\lambda f^2(x) \\
&\leq \frac{1}{w} \{ \sum_{b \in W_1} [f(bx) - f(bgx)]^2 + \sum_{b \in W_2} [f(bx) - f(gx)]^2 \\
&\quad + \sum_{b \in W_3} [f(x) - f(bgx)]^2 \} \\
&\quad + \frac{k\alpha\lambda}{w} \{ \sum_{b \in W_1} f^2(bx) + \sum_{b \in W_2} f^2(bx) + \sum_{b \in W_3} f^2(x) \}.
\end{aligned}
$$

Suppose we can show that $bx \sim gx$ for $b \in W_2$ and $x \sim bgx$ for $b \in W_3$. Then by applying Cases 1 and 2, we have $\phi_g(x) \leq \phi_a(z)$, as desired.

To see that $bx \sim gx$ for $b \in W_2$, we consider $bgx \in \delta S$ and its neighbors bx, gx are in S. By the definition of strong convexity, we have $bx \sim gx$. Also, $x \sim bgx$ for $b \in W_3$, since $bx \in \delta S$ and its neighbors x, bgx are in S.

The rest of the proof is quite similar to that of Theorem 9.9, so will be omitted.

$\square$

By taking $\alpha = 4$ in Theorem 9.9 we have

THEOREM 9.10. *Suppose S is a strongly convex subgraph in an abelian homogeneous graph Γ with edge generating set $\mathcal{K}$ consisting of k generators. Let $f : S \to \mathbb{R}$ denote an eigenfunction with associated Neumann eigenvalue λ. Then for all $x \in S, a \in \mathcal{K}$, we have*

$$[f(x) - f(ax)]^2 \leq 8k\lambda \sup_{y \in S} f^2(y).$$

9.6. Eigenvalues and diameters

The Harnack inequality in previous sections can be used to derive the following eigenvalue inequality:

THEOREM 9.11. *The Neumann eigenvalue λ_S of a convex subgraph S of an abelian homogeneous graph Γ satisfies*

$$\lambda_S \geq \frac{1}{8kD^2}$$

where k is the degree of Γ and D is the diameter of S.

PROOF. Let f denote an eigenfunction defined on δS achieving $\lambda_S = \lambda$. We can choose f such that

$$\sup_{x \in S} |f(x)| = 1 = \sup_{x \in S} f(x).$$

Let u denote a vertex with $f(u) = \max_{x \in S} f(x) = 1$ and let v denote a vertex with $f(v) < 0$. Such a v exists since $\sum_{x \in S} f(x) = 0$. We now consider a shortest path P in S joining u and v, say, $P = (u = v_0, v_1, \cdots, v_t = v)$. Since the diameter of S is D, we have $t \leq D$.

We consider

$$S = \sum_{i=0}^{t-1} [f(v_i) - f(v_{i+1})]^2.$$

By Theorem 9.10, we have

$$S \leq 8k\lambda D.$$

On the other hand, we have

$$\begin{aligned}
S &= \sum_{i=0}^{t-1} [f(v_i) - f(v_{i+1})]^2 \\
&\geq \frac{1}{D} (f(u) - f(v))^2 \\
&\geq \frac{1}{D}.
\end{aligned}$$

Therefore we obtain

$$\lambda \geq \frac{1}{8kD^2}$$

and the proof of Theorem 9.11 is complete. $\square$

CHAPTER 10

Heat kernels

10.1. The heat kernel of a graph and its induced subgraphs

Many of the most powerful techniques in spectral geometry involve heat kernels. We will consider heat kernels for a graph or an induced subgraph having Laplacian acting on functions with boundary conditions. The name "heat kernel" is due to the fact that it is a fundamental solution of the following *heat equation*:

$$\frac{\partial u}{\partial t} = -\mathcal{L}u.$$

In a way, the Laplacian is associated with the rate of dissipation of heat. To be precise, suppose for a graph G on n vertices, or for an induced subgraph S on n vertices, we consider the Laplacian $\mathcal{L}$ acting on functions with Neumann or Dirichlet boundary conditions (as defined in Chapter 8.) We can write $\mathcal{L}$ as follows:

$$\mathcal{L} = \sum_{i=0}^{n-1} \lambda_i I_i$$

where I_i is the projection onto the ith eigenfunction ϕ_i of the induced subgraph S. For any $t \geq 0$, the heat kernel H_t of G is defined to be the $n \times n$ matrix

$$
\begin{aligned}
H_t &= \sum_i e^{-\lambda_i t} I_i \\
&= e^{-t\mathcal{L}} \\
&= I - t\mathcal{L} + \frac{\epsilon^2}{2}\mathcal{L}^2 - \dots
\end{aligned}
$$

In particular,

$$H_0 = I$$

where I denotes the identity matrix.

All information about a graph is, of course, contained in its heat kernel. One of the main sources of power in using the heat kernel is the presence of the additional variable t. For example, it provides flexibility for using the maximum principle. Sometimes, it is easier to establish inequalities or equalities involving the heat kernel either for all t or for a large range of t. We can then often deduce the spectrum (or its bounds) by selecting t appropriately.

We remark that the definition for the heat kernel for graphs is exactly the parallel of the heat kernel for Riemannian manifolds (see the book of Yau and Schoen [254]). The first four sections in this chapter are adapted from [60]. Section 10.5 is based on [79].

10.2. Basic facts on heat kernels

Let S denote an induced subgraph of a graph G. Let H denote the heat kernel of S subject to Neumann or Dirichlet boundary conditions.

LEMMA 10.1. *The heat kernel H_t for a graph G with eigenfunctions ϕ_i satisfies*

$$H_t(x,y) = \sum_i e^{-\lambda_i t} \phi_i(x) \phi_i(y).$$

PROOF. The proof follows from the fact that

$$H_t = \sum_i e^{-\lambda_i t} I_i$$

and

$$I_i(x,y) = \phi_i(x)\phi_i(y).$$

$\square$

LEMMA 10.2. *For $0 \le s \le t$, we have*

$$\sum_y H_s(x,y) H_{t-s}(y,z) = H_t(x,z).$$

The proof of Lemma 10.2 follows from the definition (via matrix multiplication).

For a function $f : S \cup \delta S \to \mathbb{R}$, we consider

$$
\begin{aligned}
F(x,t) &= \sum_{y \in S \cup \delta S} H_t(x,y) f(y) \\
&= (H_t f)(x).
\end{aligned}
$$

Here we consider some useful facts about F and H_t:

LEMMA 10.3.

(i): $F(x,0) = f(x)$

(ii): *For $x \in S \cup \delta S$,*

$$\sum_{y \in S \cup \delta S} H_t(x,y) \sqrt{d_y} = \sqrt{d_x}$$

(iii): *F satisfies the heat equation:*

$$\frac{\partial F}{\partial t} = -\mathcal{L} F$$

(iv): *Under the Neumann boundary condition, we have, for the any vertex x in δS,*

$$\mathcal{L} F(x,t) = \sum_{\substack{y \\ x \sim y}} \left(\frac{F(x,t)}{\sqrt{d_x}} - \frac{F(y,t)}{\sqrt{d_y}} \right) = 0$$

For the Dirichlet boundary condition, we have, for any vertex x in δS,

$$F(x,t) = 0$$

(v):
$$\sum_{\{x,y\} \in S^*} (\frac{F(x,t)}{\sqrt{d_x}} - \frac{F(y,t)}{\sqrt{d_y}})^2 = \sum_{x \in S} F(x,t) \mathcal{L} F(x,t)$$

where S^* denotes the set of all edges with at least one endvertex in S.

PROOF. (i) is obvious, and (ii) follows by considering the all 1's function **1**:

$$\begin{aligned}
\sum_y H_t(x,y)\sqrt{d_y} &= (H_t T^{1/2} \mathbf{1})(x) \\
&= T^{1/2} \mathbf{1}(x) \\
&= \sqrt{d_x}.
\end{aligned}$$

To see (iii), we have

$$\begin{aligned}
\frac{\partial F}{\partial t} &= \frac{\partial}{\partial t} H_t f \\
&= -\mathcal{L} e^{-\mathcal{L}t} f \\
&= -\mathcal{L} F.
\end{aligned}$$

The proof of (iv) follows from the fact that all eigenfunctions f have a corresponding F which satisfies (iv).

To prove (v), we consider

$$\begin{aligned}
\sum_{x \in S} F(x,t) \mathcal{L} F(x,t) &= \sum_{x \in S} F(x,t) T^{-1/2} L T^{-1/2} F(x,t) \\
&= \sum_{x \in S} \frac{F(x,t)}{\sqrt{d_x}} \left(\sum_{\substack{y \\ \{x,y\} \in S^*}} (\frac{F(x,t)}{\sqrt{d_x}} - \frac{F(y,x)}{\sqrt{d_y}}) \right) \\
&= \sum_{\substack{y \\ \{x,y\} \in S^*}} (\frac{F(x,t)}{\sqrt{d_x}} - \frac{F(y,x)}{\sqrt{d_y}})^2
\end{aligned}$$

by using the Neumann or Dirichlet condition as described in (iv).   □

LEMMA 10.4. *For all $x, y \in S \cup \delta S$, we have*

$$H_t(x,y) \geq 0.$$

PROOF. The matrix $\mathcal{A} = I - \mathcal{L}$ has all entries non-negative. Therefore $e^{t\mathcal{A}}$ has all non-negative entries. Since

$$H_t = e^{-t} e^{t\mathcal{A}},$$

all entries of H_t are non-negative. The lemma is proved.   □

10.3. An eigenvalue inequality

In this section, we will establish an inequality for lower bounding the eigenvalue λ_1 of a graph or a subgraph with boundary. The main method here is by repeatedly using the heat kernel and using various properties of the heat kernel.

Suppose we have a function $f : S \cup \delta S \to \mathbb{R}$. We consider

$$(10.1) \qquad g(x,t) = \sum_{y \in S} H_t(x,y)\sqrt{d_x d_y}\left(\frac{f(y)}{\sqrt{d_y}} - \frac{F(x,t)}{\sqrt{d_x}}\right)^2.$$

where $F(x,t) = \sum_y H_t(x,y)f(y)$. Therefore, we have

$$(10.2) \qquad g(x,t) = \sum_{y \in S} H_t(x,y)\sqrt{d_x/d_y}f^2(y) - F^2(x,t)$$

By summing over x, we have

$$\sum_{x \in S} g(x,t) = \sum_{x \in S}\sum_{y \in S} H_t(x,y)\sqrt{d_x/d_y}f^2(y) - \sum_x F^2(x,t).$$

Using Lemma 10.3 (ii), (iv) and (v), we obtain

$$
\begin{aligned}
\sum_{x \in S} g(x,t) &= \sum_{y \in S} f^2(y) - \sum_x F^2(x,t) \\
&= -\int_0^t \frac{d}{ds}\sum_{x \in S} F^2(x,s)ds \\
&= -2\int_0^t \sum_{x \in S} F(x,s)\frac{d}{ds}F(x,s)ds \\
&= 2\int_0^t \sum_{x \in S} F(x,s)\mathcal{L}F(x,s)ds \\
(10.3) \qquad &= 2\int_0^t \sum_{\{x,y\} \in S^*}\left(\frac{F(x,t)}{\sqrt{d_x}} - \frac{F(y,t)}{\sqrt{d_y}}\right)^2 ds.
\end{aligned}
$$

We claim that for $t \ge 0$, we have

Fact 1: $\displaystyle\sum_{\{x,y\} \in S^*}\left(\frac{F(x,t)}{\sqrt{d_x}} - \frac{F(y,t)}{\sqrt{d_y}}\right)^2 \le \sum_{\{x,y\} \in S^*}\left(\frac{f(x)}{\sqrt{d_x}} - \frac{f(y)}{\sqrt{d_y}}\right)^2.$

To see this, we consider

$$\frac{d}{dt}\sum_{\{x,y\} \in S^*}\left(\frac{F(x,t)}{\sqrt{d_x}} - \frac{F(y,t)}{\sqrt{d_y}}\right)^2$$

$$(10.4) \qquad = 2\sum_{\{x,y\} \in S^*}(\frac{F(x,t)}{\sqrt{d_x}} - \frac{F(y,t)}{\sqrt{d_y}})(\frac{d}{dt}\frac{F(x,t)}{\sqrt{d_x}} - \frac{d}{dt}\frac{F(y,t)}{\sqrt{d_y}}).$$

For the Laplacian with Neumann boundary condition, the above expression is equal to

$$2 \sum_{x \in S \cup \delta S} \frac{d}{dt} \frac{F(x,t)}{\sqrt{d_x}} \left(\frac{F(x,t)}{\sqrt{d_x}} - \frac{F(y,t)}{\sqrt{d_y}} \right)$$

$$= 2 \sum_{x \in S} \frac{d}{dt} \frac{F(x,t)}{\sqrt{d_x}} \left(\frac{F(x,t)}{\sqrt{d_x}} - \frac{F(y,t)}{\sqrt{d_y}} \right)$$

$$= 2 \sum_{x \in S} \frac{d}{dt} \frac{F(x,t)}{\sqrt{d_x}} \mathcal{L} F(x,t).$$

For the Laplacian with Dirichlet boundary condition, (10.4) is equal to

$$2 \sum_{x \in S \cup \delta S} \frac{F(x,t)}{\sqrt{d_x}} \sum_{\substack{y \\ \{x,y\} \in S^*}} \left(\frac{d}{dt} \frac{F(x,t)}{\sqrt{d_x}} - \frac{d}{dt} \frac{F(y,t)}{\sqrt{d_y}} \right)$$

$$= 2 \sum_{x \in S} \frac{F(x,t)}{\sqrt{d_x}} \sum_{\substack{y \\ \{x,y\} \in S^*}} \left(\frac{d}{dt} \frac{F(x,t)}{\sqrt{d_x}} - \frac{d}{dt} \frac{F(y,t)}{\sqrt{d_y}} \right)$$

$$= 2 \sum_{x \in S} F(x,t) \mathcal{L} \frac{d}{dt} F(x,t).$$

In either case, (10.4) is equal to

$$2 \sum_{x \in S} \frac{d}{dt} F(x,t) \mathcal{L} F(x,t)$$

$$= -2 \sum_{x \in S} \frac{d}{dt} F(x,t) \cdot \frac{d}{dt} F(x,t)$$

$$= -2 \sum_{x \in S} (\frac{d}{dt} F(x,t))^2$$

$$\leq 0.$$

Therefore

$$\sum_{\{x,y\} \in S^*} \left(\frac{F(x,t)}{\sqrt{d_x}} - \frac{F(y,t)}{\sqrt{d_y}} \right)^2 \leq \sum_{\{x,y\} \in S^*} \left(\frac{F(x,0)}{\sqrt{d_x}} - \frac{F(y,0)}{\sqrt{d_y}} \right)^2$$

$$= \sum_{\{x,y\} \in S^*} \left(\frac{f(x)}{\sqrt{d_x}} - \frac{f(y)}{\sqrt{d_y}} \right)^2,$$

and Fact 1 is proved.

Substituting the inequality of Fact 1 into (10.3), we get

$$\sum_{x \in S} g(x,t) = 2 \int_0^t \sum_{\{x,y\} \in S^*} \left(\frac{F(x,s)}{\sqrt{d_x}} - \frac{F(y,s)}{\sqrt{d_y}} \right)^2 ds$$

(10.5)

$$\leq 2t \sum_{\{x,y\} \in S^*} \left(\frac{f(x)}{\sqrt{d_x}} - \frac{f(y)}{\sqrt{d_y}} \right)^2.$$

In the other direction, we consider the lower bound:

$$\sum_{x \in S} g(x,t) = \sum_{x \in S} \sum_{y \in S} H_t(x,y) \sqrt{d_x d_y} \left(\frac{f(y)}{\sqrt{d_y}} - \frac{F(x,t)}{\sqrt{d_x}} \right)^2$$

$$\geq \sum_{x \in S} \left(\inf_{y \in S} H_t(x,y) \sqrt{\frac{d_x}{d_y}} \right) \sum_{y \in S} \left(\frac{f(y)}{\sqrt{d_y}} - \frac{F(x,t)}{\sqrt{d_x}} \right)^2 d_y$$

$$(10.6) \qquad \geq \left(\sum_{x \in S} \inf_{y \in S} \frac{H_t(x,y)\sqrt{d_x}}{\sqrt{d_y}} \right) \inf_{c \in \mathbb{R}} \sum_{y} \left(\frac{f(y)}{d_y} - c \right)^2 d_y.$$

Combining (10.5) and (10.6), we have

$$\sup_{c} \frac{\displaystyle\sum_{\substack{y \\ \{x,y\} \in S^*}} \left(\frac{f(x)}{\sqrt{d_x}} - \frac{f(y)}{\sqrt{d_y}} \right)^2}{\displaystyle\sum_{y \in S} \left(\frac{f(y)}{\sqrt{d_y}} - c \right)^2 d_y} \geq \frac{1}{2t} \sum_{x \in S} \inf_{y \in S} \frac{H_t(x,y)\sqrt{d_x}}{\sqrt{d_y}}.$$

Therefore we have proved the following:

THEOREM 10.5.
(i) The Neumann eigenvalue λ_1 for an induced subgraph S with heat kernel H_t satisfies

$$(10.7) \qquad\qquad \lambda_1 \geq \frac{1}{2t} \sum_{x \in S} \inf_{y \in S} \frac{H_t(x,y)\sqrt{d_x}}{\sqrt{d_y}}.$$

(ii) The Dirichlet eigenvalue λ_1' for an induced subgraph S with heat kernel H_t' satisfies

$$(10.8) \qquad\qquad \lambda_1' \geq \frac{1}{2t} \sum_{x \in S} \inf_{y \in S} \frac{H_t'(x,y)\sqrt{d_x}}{\sqrt{d_y}}.$$

10.4. Heat kernel lower bounds

Theorem 10.5 can be used to establish eigenvalue lower bounds if we can effectively find good lower bounds for the heat kernels. Here, we discuss two examples. The first example is a direct and simple way of using the inequality (10.7). The second example is to bound the kernel for a "convex subgraph" by relating it to an associated continuous heat kernel for an appropriate Riemannian manifold.

We consider the special case of a graph G with no boundary. In addition, suppose the graph G has a "covering" vertex x_0 with the property that x_0 is adjacent to every other vertex y in G. The degree of x_0 is $n-1$ where G has n vertices. We will apply (10.7) with $t \to 0$. Since

$$H_t = I - t\mathcal{L} + O(t^2),$$

it follows from (10.7) that

$$\lambda_1 \geq \frac{1}{2t} \sum_{x \in S} \inf_{y \in S} \frac{H_t(x,y)\sqrt{d_x}}{\sqrt{d_y}}$$

$$= \frac{1}{2t}\left(\inf_{y \neq x_0} \frac{t}{d_y} + O(t^2) \right)$$

$$= \frac{1}{2t}\left(\frac{t}{\delta_2} \right)$$

where δ_2 denotes the second largest degree in G. Thus, we have

(10.9)
$$\lambda_1 \geq \frac{1}{2\delta_2}.$$

For example, for $G = P_3$, the path with 3 vertices, it is true that $\lambda_0 = 0, \lambda_1 = 1$, and $\lambda_2 = 2$, while our estimate in (10.9) gives $\lambda_1 \geq 1/2$. Applying this to $G = K_n$, the complete graph on n vertices, yields

$$\lambda_1 \geq \frac{n}{2(n-1)}$$

while the true value is $\lambda_1 = n/(n-1)$ (again off by a factor of 2). In fact, as pointed out by L. Lovász, $\lambda_1 \geq 1/\delta_2$ follows by directly using the Rayleigh quotient.

For the remainder of this section, we will restrict ourselves to special subgraphs of homogeneous graphs that are embedded in Riemannian manifolds. Such a restriction will allow us to derive eigenvalue bounds for graphs using the known results on eigenvalues of Riemannian manifolds. Also, the restricted classes of graphs still include many families of graphs which arise in various applications in enumeration and sampling. We remark that in Chapter 9, we derive eigenvalue lower bounds for subgraphs of homogeneous graphs with stronger convexity conditions by using a Harnack inequality. Both the definitions and the methods are different here. Roughly speaking, our goal here is to use the (continuous) heat kernels of the manifolds with convex boundary for deriving as the lower bound function for the (discrete) heat kernel.

Suppose Γ is a homogeneous graph with associated group $\mathcal{H}$. Here we assume that the edge generating set $\mathcal{K} \subset \mathcal{H}$ is symmetric, i.e., $g \in \mathcal{K}$ if and only if $g^{-1} \in \mathcal{K}$.

Suppose vertices of Γ can be embedded into a manifold $\mathcal{M}$ with a measure μ such that

$$\mu(x, gx) = \mu(y, gy)$$

for any $g \in \mathcal{K}$ and $x, y \in V(\Gamma)$. In addition, if

$$\mu(x, gx) = \mu(y, g'y)$$

for any $g, g' \in \mathcal{K}$ and $x, y \in V(\Gamma)$ then Γ is called a *lattice* graph. An induced subgraph on a subset S of vertices of a lattice graph Γ is said to be a convex subgraph if there is a submanifold M with convex boundary such that S consists of all vertices of Γ contained in M, i.e.,

$$S = M \cap V(\Gamma).$$

Furthermore, we require that for any vertex x, the *Voronoi* region $R_x = \{y : \mu(y, x) < \mu(y, z)$ for all $z \in \Gamma \cap \mathcal{M}\}$ is contained in M.

EXAMPLE 10.6. As in Example 9.3., we consider a homogeneous graph Γ with vertex set consisting of all $m \times n$ matrices with integral (possibly negative) entries having column sums $c_1, \ldots, c_n$, and row sums $r_1, \ldots r_m$. Two vertices x and y are adjacent if they differ at four entries in some submatrix determined by two columns i, j and two rows k, m satisfying

$$x_{ik} = y_{ik} + 1, x_{jk} = y_{jk} - 1, x_{im} = y_{im} - 1, x_{jm} = y_{jm} + 1$$

It is easy to see that Γ is a homogeneous graph with edge generating set consisting of all 2×2 submatrices $\begin{pmatrix} 1 & -1 \\ -1 & 1 \end{pmatrix}$. Obviously, Γ can be viewed as being embedded in the mn-dimensional Euclidean space $\mathcal{M} = \mathbb{R}^{mn}$. In fact, Γ is embedded in a $(mn - m - n + 1)$-dimensional subspace of $\mathcal{M}$. We consider the set S of all $m \times n$ matrices with non-negative integral entries with given row and column sums. Suppose we choose the submanifold M, determined by

$$\sum_j x_{ij} = r_i$$

$$\sum_i x_{ij} = c_i$$

$$x_{ij} > -\frac{1}{2}.$$

It is easy to verify that S is a convex subgraph of the lattice graph Γ.

Here we state the main theorem for bounding the eigenvalue λ_1 of a convex subgraph.

THEOREM 10.7. *Let S denote a convex subgraph of a lattice graph and suppose S is embedded into a d-dimensional manifold M with a convex boundary and a distance function μ. Let $\mathcal{K}$ denote the set of edge generators and suppose $\epsilon = \min\{\mu(x, gx) : g \in \mathcal{K}\}$. In the first order approximation of the discrete Laplacian $\mathcal{L}$ by the continuous Laplace operator,*

$$(10.10) \qquad -\frac{2d}{\epsilon^2}\mathcal{L} \sim \frac{2d}{|\mathcal{K}|} \sum_{g \in \mathcal{K}^*} (\frac{\epsilon}{\mu(x, gx)})^2 \frac{\partial^2}{\partial g^2}$$

$$= \sum_{i,j} a_{i,j} \frac{\partial^2}{\partial x_i \, \partial x_j}$$

where $\mathcal{K}^*$ consists of exactly one of a and a^{-1} for all $a \in \mathcal{K}$. Suppose that

$$C_1 I \leq (a_{i,j}) \leq C_2 I$$

where I is the identity matrix. Then the Neumann eigenvalue λ_1 of S satisfies the following inequality:

$$\lambda_1 \geq \frac{c_0 \, r \, \epsilon^2}{d^2 D(M)^2}$$

where

$$r = \frac{U \, |S|}{\text{vol } M},$$

U denotes the volume of the Voronoi region, and c_0 is an absolute constant satisfying

$$c_0 \leq C_0 \min\{C_1, C_2^{-1}\}$$

for some absolute constant C_0.

The proof of Theorem 10.7 is rather long and complicated. We will sketch the major ideas here; the details can be found in [60]. There are three major parts.

Part 1: Lower bounds for the continuous heat kernel.

Let $h(t, x, y)$ denote the heat kernel of M and let $u(t, x) = h(t, x, y)$ satisfy the heat equation

$$(\Delta - \frac{\partial}{\partial t})u(t, x) = 0$$

with the Neumann boundary condition

$$\frac{\partial}{\partial \nu}u(t, x) = 0$$

for any boundary point x.

Here the Laplace operator Δ is taken to be

$$\Delta = \frac{\epsilon^2}{|\mathcal{K}|} \sum_{g \in \mathcal{K}} \frac{\partial^2}{\partial g^2}.$$

Also, we assume that

$$\mu(x, gx) = \epsilon$$

for all $x \in V(\Gamma)$ and $g \in \mathcal{K}$.

Li and Yau established the following lower bound for H in their seminal paper [187]:

Theorem: *Let M denote a d-dimensional compact manifold with boundary ∂M. Suppose the Ricci curvature of M is nonnegative, and if $\partial M \neq \emptyset$, we assume further that ∂M is convex. Then the fundamental solution of the heat equation with the Neumann boundary condition satisfies*

$$h(t, x, y) \geq C^{-1}(\epsilon)(\text{vol}(B_x(\sqrt{t})))^{-1}exp\frac{-\mu(x, y)}{(4 - \epsilon)t}$$

for some constant $C(\epsilon)$ depending on $\epsilon > 0$ and d such that $C(\epsilon) \to \infty$ as $\epsilon \to 0$. (Here, vol($\cdot$) denotes volume and $B_x(r)$ denotes the intersection of M with the ball of radius r centered at x.)

However, the above version of the usual estimates for the heat kernel cannot be directly used for our purposes here since the constant C is exponentially small depending on d. A more careful analysis of the heat kernel is needed. To lower-bound the discrete heat kernel, we will use the following lower bound estimates for the (continuous) heat kernel (a proof can be found in [60]).

For any $\alpha > 0$, and $\sigma \geq cd\alpha$,

(10.11) $\qquad h(t,x,y) \geq \dfrac{(1+\alpha)^{-d}}{4B_x(\sqrt{\sigma}t)} exp \dfrac{-(1+\alpha)\mu^2(x,y)}{\alpha t}.$

Suppose we choose $\alpha = \frac{1}{d}$, and $c_2 t_0 = dD^2(M)$, where $c_2 = 2\epsilon^2/d$ and $D(M)$ denotes the diameter of M. (We may assume $D(M) \geq 1$.) By using (10.11) we have

$$h(c_2 t_0, x, y) \geq \frac{c'}{\text{vol } M}.$$

Part 2: Approximate the discrete heat kernel by the continuous heat kernel.

We define the following function $k(t,x,y)$ which will serve as a lower bound for the heat kernel of the graph.

$$k(t,x,y) = c_1 \int_M h(c_2\, t, x - z, y)\varphi(z)dz$$

where φ is a bell-shaped function, for example, a modified Gaussian function $exp(-c'|z/\epsilon|^2)$ with compact support, say, $\{|z| < \epsilon/4\}$, and which satisfies

(10.12) $\qquad c_1 \displaystyle\int \varphi(z)dz = c_3\ (c_4\epsilon)^d$

where c_3 and c_4 are chosen so that the above quantity is within a constant factor of the volume of the Voronoi region R_x. So, $k(t,x,y)$ can be approximated by $h(c_2 t, x, y)U$ or

(10.13) $\qquad \displaystyle\int_{R_x} h(c_2 t, z, y)dz$

when t is not too small (using the gradient estimates of h which are also proved in [60])). Here U denotes the maximum over x of the volume of R_x.

From (10.12), we have

$$
\begin{aligned}
k(t_0 - \epsilon', x, y) &= c_1 \int_M h(c_2\,(t_0 - \epsilon'), x - z, y)\varphi(z)dz \\
&\geq \frac{c_1 c_6}{\text{vol } M} \int_M \varphi(z)dz \\
&\geq \frac{c_7\, U}{\text{vol } M}.
\end{aligned}
$$

where $c_7 \leq c' \min\{C_1, C_2^{-1}\}$ and the Harnack inequalities in the continuous case are used.

We note that the c's (with the exception of c_2, which is a scaling factor) denote some appropriate absolute constants. In the next section, we will give an example of computing the c's in the eigenvalue computation of Example 10.7.

In order to show that $k(t,x,y)$ is indeed a lower bound for $H_t(x,y)$, a proof is needed. There are several sets of sufficient conditions for establishing lower bounds of $H_t(x, y)$. The detailed proofs in [**60**] are somewhat long and will not be reproduced here. Using this, we then have

$$H_t(x, y) \geq \frac{c_7 U}{\text{vol } M}.$$

Part 3: Eigenvalue lower bound.

Now we can combine all the estimates and use Theorem 10.5. Hence,

$$
\begin{aligned}
\lambda \quad &\geq \quad \frac{1}{2t} \sum_{x \in S} \inf_{y \in S} H_t(x, y) \\
&\geq \quad \frac{c_7 \, U \, |S|}{t \, \text{vol } M} \\
&\geq \quad \frac{c_7 \, \epsilon^2 \, r}{d^2 \, D(M)^2}
\end{aligned}
$$

where U denotes the volume of the Voronoi region and r denotes the ratio $U|S|/\text{vol } M$. This completes the sketch of the proof for Theorem 10.8.

To get a simpler lower bound for λ, we note that the diameter $D(S)$ of the convex subgraph S and the diameter of the manifold are related by

(10.14) $$D(M) \leq \epsilon \, D(S).$$

Therefore, we have the following:

COROLLARY 10.8. *Let S denote a convex subgraph of a simple lattice graph and suppose S is embedded into a d-dimensional manifold M with a convex boundary. Then the Neumann eigenvalue λ_1 of S satisfies the following inequality:*

$$\lambda_1 \geq \frac{c_0 \, r}{d^2 D^2(S)}$$

where

$$r = \frac{U \, |S|}{\text{vol } M},$$

$D(S)$ denotes the (graph) diameter of S, $\mathcal{K}$ denotes the set of edge generators, and c_0 is an absolute constant depending only on the simple lattice graph.

REMARK 10.9. For a polytope in $\mathbb{R}^d$, we can rescale and choose the lattice points to be dense enough to approximate the volume of the polytope. For example, if we have

(10.15) $$C \, \epsilon \leq D_1(M)/d$$

where D_1 denotes the diameter of M measured by the L_1 norm and C is some absolute constant, then the number of lattice points provides a good approximation for the volume of the polytope. This implies that $r \geq c$ for some constant c. The above inequality (10.15) can be replaced by a slightly simpler inequality:

$$C' \, d \leq D(S)$$

for some constant C'. These facts are useful for approximation algorithms for the volume of a convex body, which we discuss in the next section.

REMARK 10.10. There are many graphs G that can be embedded in a lattice graph in such a way that the diameter of G satisfies

$$D(G) \sim \frac{\sqrt{d}\, D(M)}{\epsilon}.$$

For such graphs, Theorem 10.7 implies a somewhat stronger result:

$$\lambda \geq \frac{c_0 r}{d^2 D(G)^2}$$

where r is as defined in Theorem 10.7.

10.5. Matrices with given row and column sums

One of the classical problems in enumerative combinatorics is to count (or estimate) the number of matrices with nonnegative entries having given row and column sums (as described in Example 10.7). This problem arises in a variety of applications, such as graph enumeration, goodness of fit tests in statistics, enumeration of permutations by descents, describing tensor product decompositions, counting double cosets, etc. In particular, in statistics it is often referred to as the "contingency table" problem and has a long history. It seems to be a particularly difficult problem to obtain good estimates for the number of such tables with given row and column sums. In order to attack this problem, a standard technique depends on rapidly generating random tables with nearly equal probability.

In this section, we will use the theorems in the preceding sections to estimate the eigenvalue λ_1 for the convex subgraph with vertices corresponding to matrices with given row and column sums. By using these bounds we can derive convergence bounds for Neumann random walks on this subgraph. This sections is based on [**79**].

Given integer vectors $\bar{r} = (r_1, \ldots, r_m)$, $\bar{c} = (c_1, \ldots, c_n)$ with $r_i, c_j \geq 0$ and $\sum_i r_i = \sum_j c_j$, we consider the space of all $m \times n$ arrays T with non-negative integer entries satisfying with the property that

$$\sum_j T(i,j) = r_i, \qquad 1 \leq i \leq m$$

$$\sum_i T(i,j) = c_j, \qquad 1 \leq j \leq n .$$

Let us denote by $\mathcal{T} = \mathcal{T}(\bar{r}, \bar{c})$ the set of all such arrays.

We will show that for the Neumann walk on the space of tables $\mathcal{T}(\bar{r}, \bar{c})$ where

$$\min \left\{ \min_i \frac{r_i}{n}, \ \min_j \frac{c_j}{m} \right\} > c(m-1)^{3/2}(n-1)^{3/2}$$

we have

(10.16) $$\lambda_1 > \left[3200 e^{1/c}(m-1)^2(n-1)^2 \min \left\{ \sum_i r_i^2, \sum_j c_j^2 \right\} \right]^{-1} .$$

We first need to place our contingency table problem into the framework of the preceding section. The manifold $\mathcal{M}$ will consist of all real mn-tuples $\bar{x} = (x_{11}, x_{12}, \ldots, x_{mn})$ satisfying

$$\sum_j x_{ij} = r_i, \quad \sum_i x_{ij} = c_j .$$

Since $\sum_i r_i = \sum_j c_j$ then

$$\dim \mathcal{M} = N := (m-1)(n-1) .$$

The graph Γ has as vertices all the integer points in $\mathcal{M}$, i.e., all $\bar{x}$ with all $x_{ij} \in \mathbb{Z}$. The edge generating set $\mathcal{K}$ consists of all the basic moves described above. Hence, $|\mathcal{K}| = \binom{m}{2}\binom{n}{2}$. The set S will just be $\mathcal{T} = \mathcal{T}(\bar{r}, \bar{c})$, the set of all $T \in \Gamma$ with all entries nonnegative. Thus,

$$S = \cap_{i,j}\{T \in \Gamma : x_{ij} \geq 0\} .$$

Similarly, the manifold $M \subset \mathcal{M}$ is defined by

$$M = \cap_{i,j}\{\bar{x} \in \mathcal{M} : x_{ij} > -1/2\} .$$

It is clear that M is an N-dimensional convex polytope and $S = \mathcal{T}$ is the set of all lattice points in M, and consequently convex in the sense needed for (10.10). It is easy to see that $\mathcal{T}$ is connected by the basic moves generated by $\mathcal{K}$, and that each edge of Γ has length 2. Our next problem is to deal with the term $\frac{|S| \operatorname{vol} U}{\operatorname{vol} M}$ in (10.10). In particular, we would like to show this is close to 1, provided that r_i and c_j are not too small. To do this we need the following two results.

Claim 1.

Suppose $L \subset \mathbb{R}^N$ is a lattice generated by vectors $v_1, \ldots, v_N$. Then the covering radius of L is at most $R := \frac{1}{2}\left(\sum_{i=1}^{N} \|v_i\|^2\right)^{1/2}$.

PROOF. The assertion clearly holds for $N = 1$. Assume it holds for all dimensions less than N. It is enough to prove that any point $\bar{x} = (x_1, \ldots, x_N)$ in the fundamental domain generated by the v_i is at most a distance of R from some v_i. Let $\bar{x}_0$ be the projection of $\bar{x}$ on either the hyperplane generated by $v_1, \ldots, v_{N-1}$, or a translate of the hyperplane by v_N, whichever is closer (these are two bounding hyperplanes of the fundamental domain). Thus, $d(\bar{x}, \bar{x}_0) \leq \frac{1}{2}\|v_N\|$. By the induction hypothesis,

$$d(\bar{x}_0, v_j) \leq \frac{1}{2}\left(\sum_{i=1}^{N-1} \|v_i\|^2\right)^{1/2} \quad \text{for some } j < N .$$

Hence,

$$d(\bar{x}, v_j) \leq \left(\frac{1}{4}\|v_N\|^2 + \frac{1}{4}\sum_{i=1}^{N-1} \|v_i\|^2\right)^{1/2} = R$$

as claimed. $\qquad \square$

Claim 2. If M is convex and contains an open ball $B(cRN)$ of radius cRN, with $cN \geq 2$, then

(10.17)
$$4^{-1/c} \leq \frac{|S| \operatorname{vol} U}{\operatorname{vol} M} \leq 1$$

where $v_1, \ldots v_N$ generate Γ, and $R = \frac{1}{2} \left(\sum_{v=1}^{N} \|v_i\|^2 \right)^{1/2}$.

PROOF. Consider a contracted copy $(1 - \delta)M$ of M with the origin centered about the center of the ball $B(cRN) \subset M$. Let L be any bounding hyperplane of M, and let $(1 - \delta)L$ be the corresponding contracted copy of L.

Let $x \in (1 - \delta)M$ and suppose there exists a lattice point $y \in V(\Gamma) \setminus S$ such that x is contained in the Voronoi region for y. Hence,

$$R > d(x, y) \geq \gamma \geq c\delta RN .$$

However, this is a contradiction if we take $\delta = \frac{1}{cN}$.

Consequently, for the choice $\delta = \frac{1}{cN}$ we must have x in the closure of the Voronoi region of some lattice point in S. Therefore,

$$|S| \operatorname{vol} U \geq \operatorname{vol}(1 - \delta)M$$

i.e.,

$$\frac{|S| \operatorname{vol} U}{\operatorname{vol} M} \geq \left(1 - \frac{1}{cN} \right)^N \geq 4^{-1/c}$$

if $cN \geq 2$. Claim 2 is proved.   □

In order to apply the result in Claim 2, we must find a large ball in M. Let s_0 denote the smallest *line sum average*, i.e.,

$$s_0 = \min \left(\min_i \frac{r_i}{n}, \ \min_j \frac{c_j}{m} \right) .$$

We begin constructing an element $T_0 \in M$ recursively as follows. Suppose without loss of generality that

$$s_0 = \frac{r_1}{m} .$$

Then, in T_0, set all elements of the first row equal to s_0, and subtract s_0 from each value c_j to form $c'_j = c_j - s_0$, $1 \leq j \leq n$. Now, to complete T_0, we are reduced to forming an $(m - 1)$ by n table T'_0 with row sums $r_2, \ldots r_m$ and column sums $c'_1, \ldots, c'_n$. The key point here is that all the line sum averages for T'_0 are still at least as large as s_0. Hence, continuing this process we can eventually construct a table T_0 (with rational entries) having least entry equal to s_0. Consequently there is a ball $B(s_0)$ of radius s_0 centered at $T_0 \in M$ which is contained in M (since to leave M, some entry must become negative). Therefore, if we assume $s_0 > cN^{3/2}$ then by Theorem 10.7 and (10.17),

(10.18)
$$\lambda_1 \geq \frac{4c_0 4^{-1/c}}{N^2 (\operatorname{diam} M)^2}$$

for some absolute constant $c_0 > 0$ where $cN \geq 2$ (since for tables, all the generators have length 2, so that $R \leq N^{1/2}$).

At the end of this section, we illustrate how a specific value can be derived here for c_0 (as well as in several other cases of interest, as well). In particular, for contingency tables, we can take $c_0 = 1/800$.

Since

$$\text{diam } M < 2 \min \left\{ \left(\sum_i r_i^2 \right)^{1/2}, \left(\sum_j c_j^2 \right)^{1/2} \right\}$$

then (10.18) can be written as follows:

For the natural Neumann walk P on the space of tables $\mathcal{T}(\bar{r}, \bar{c})$ where

$$\min \left\{ \min_i \frac{r_i}{n}, \min_j \frac{c_j}{m} \right\} > c(m-1)^{3/2}(n-1)^{3/2}$$

we have

$$(10.19) \qquad \lambda_1 > \left[3200 e^{1/c} (m-1)^2 (n-1)^2 \min \left\{ \sum_i r_i^2, \sum_j c_j^2 \right\} \right]^{-1}.$$

To convert the estimate in (10.16) to an estimate for the rate of convergence of P to its stationary (uniform) distribution π, we consider

$$\Delta(t) < (1 - \lambda_1)^t \frac{\text{vol } S}{\min_x \deg_\Gamma x}$$

$$(10.20)$$

$$\leq e^{-\lambda_1 t} \frac{\text{vol } S}{\deg \Gamma} = e^{-\lambda t} |S|.$$

and λ_1 is the least nontrivial Neumann eigenvalue of Laplacian of S.

Thus, if

$$(10.21) \qquad t > \frac{2}{\lambda_1} \ln \frac{|\mathcal{T}|}{\epsilon}$$

then $\Delta(t) < \epsilon$. Note that

$$|\mathcal{T}| \leq \min \left\{ \prod_i r_i^n, \prod_j c_j^m \right\}.$$

Thus, by (10.16) and (10.21), if

$$(10.22)$$

$$t > 6400 e^{1/c} m^2 n^2 \min \left\{ \sum_i r_i^2, \sum_j c_j^2 \right\} \left(\ln \frac{1}{\epsilon} + \min \left\{ n \sum_i \ln r_i, m \sum_j \ln c_j \right\} \right)$$

then $\Delta(t) < \epsilon$, provided

$$\min \left\{ \min_i \frac{r_i}{n}, \min_j \frac{c_j}{n} \right\} > c(m-1)^{3/2}(n-1)^{3/2}.$$

It remains to bound the constant c_0 occurring in (10.10). Briefly, as in Theorem 10.7, we have

$$-\frac{2d}{l^2}\mathcal{L}_s \quad \sim \quad \frac{2d}{|\mathcal{K}|}\sum_{g\in\mathcal{K}^*}\left(\frac{\mu(x,gx)}{l}\right)^2\frac{\partial^2}{\partial g^2}$$

$$= \quad \sum_{i,j}a_{ij}\frac{\partial^2}{\partial x_i\partial x_j}$$

where $d := \dim M$, $\ell := \min_{g\in\mathcal{K}}\mu(x,gx)$, μ denotes (Euclidean) length, and $\mathcal{K}^* \subset \mathcal{K}$ consists of exactly one element from each pair $\{g,g^{-1}\}$, $g \in \mathcal{K}$.

Suppose C_1 and C_2 are constants so that

$$C_1 I \le (a_{ij}) \le C_2 I$$

where I is the identity operator on M, and $X \le Y$ means that the operator $Y - X$ is non-negative definite. In particular, we can take for C_1 and C_2 the least and greatest eigenvalues, respectively, of (a_{ij}) restricted to M. Now, from the arguments in [**60**], it follows that when $\mathcal{M}$ is Euclidean then the constant c_0 in (10.10) can be taken to be

$$c_0 = \frac{1}{100}\min(C_1, C_2^{-1})\,.$$

Thus, to determine c_0 in various applications, our job becomes that of bounding the eigenvalues of the corresponding matrix (a_{ij}).

First, we consider $m \times n$ contingency tables. With each edge generator $g = x_{ij} - x_{i'j} - x_{ij'} + x_{i'j'}$ we consider $\frac{\partial^2}{\partial g^2}$ in terms of the x's.

Expanding, we have

$$\frac{\partial^2}{\partial g^2} = \frac{\partial^2}{\partial x_{ij}^2} + \frac{\partial^2}{\partial x_{i'j}^2} + \frac{\partial^2}{\partial x_{ij'}^2} + \frac{\partial^2}{\partial x_{i'j'}^2} - 2\frac{\partial^2}{\partial x_{ij}\partial x_{i'j}}$$

$$-2\frac{\partial^2}{\partial x_{ij}dx_{ij'}} - 2\frac{\partial^2}{\partial x_{i'j'}dx_{i'j}} - 2\frac{\partial^2}{\partial x_{i'j'}dx_{ij'}}$$

$$+2\frac{\partial^2}{\partial x_{ij}\partial x_{i'j'}} + 2\frac{\partial^2}{\partial x_{i'j}dx_{ij'}}\,.$$

We can abbreviate this in matrix form as

	x_{ij}	$x_{i'j}$	$x_{ij'}$	$x_{i'j'}$
x_{ij}	1	-1	-1	1
$x_{i'j}$	-1	1	1	-1
$x_{ij'}$	-1	1	1	-1
$x_{i'j'}$	1	-1	-1	1

We need to consider the operator

$$\sum_{g\in\mathcal{K}^*}\frac{\partial^2}{\partial g^2}\,.$$

The corresponding matrix Q has the following coefficient values for its various entries:

Entry	Coefficient
(x_{ij}, x_{ij})	$(m-1)(n-1)$
$(x_{ij}, x_{i'j})$	$-(m-1)$
$(x_{ij}, x_{ij'})$	$-(m-1)$
$(x_{ij}, x_{i'j'})$	1

Thus, Q has two distinct eigenvalues: one is mn with multiplicity $(m-1)(n-1)$, and the other is 0 with multiplicity $m+n-1$. Now, $\dim M = (m-1)(n-1)$ and the operator corresponding to Q when restricted to M has all eigenvalues equal to mn. So the matrix (a_{ij}) has all eigenvalues equal to $\frac{2mn(m-1)(n-1)}{\binom{m}{2}\binom{n}{2}} = 8$, and consequently we can take $C_1 = C_2 = 8$, and $c_0 = 1/800$. This completes the computation of c_0 in the lower bound of λ_1 for the contingency table problem.

In [79], using similar methods, eigenvalue lower bounds are derived for a number of problems including restricted contingency tables, symmetric tables, compositions of an integers, and so-called knapsack solutions.

10.6. Random walks and the heat kernel

In the previous sections, we discussed methods for bounding eigenvalues using the heat kernel. As we will see here, heat kernels also play a crucial role in bounding the rate of convergence for random walks in a direct way. We will consider the relative pointwise distance (as defined in Section 1.5) to the stationary distribution for a random walk with an associated weighted graph G.

THEOREM 10.11. *Suppose a random walk is associated with a weighted graph G with heat kernel H_t. The relative pointwise distance of the random walk of t steps to the stationary distribution is bounded above by*

$$\Delta(t) \leq |\max_{x,y} H_t(x,y) \frac{\text{vol } G}{\sqrt{d_x d_y}} - 1|$$

provided $\lambda_{n-1} - 1 \leq 1 - \lambda_1$.

PROOF. Let P denote the transition probability matrix of the random walk. We note that the convergence of the random walk P^s after s steps is related to the heat kernel H_s as follows:

$$(10.23) \qquad \|P^s - T^{-1/2}I_0T^{1/2}\| = \|T^{1/2}P^sT^{-1/2} - I_0\|$$
$$= \|\sum_{i \neq 0}(1-\lambda_i)^sI_i\|$$
$$\leq e^{-\lambda_1 s}$$
$$(10.24) \qquad = \|H_s - I_0\|$$

where we recall that I_i is the projection onto the eigenspace generated by the i-th eigenfunction.

After s steps, the relative pointwise distance of P^s to the stationary distribution $\pi(x)$ is given by

$$\Delta(s) = \max_{x,y} \frac{|P^s(y,x) - \pi(x)|}{\pi(x)}.$$

Let ψ_x be defined by

$$\psi_x(y) = \begin{cases} 1 & \text{if } y = x, \\ 0 & \text{otherwise.} \end{cases}$$

We have

$$
\begin{aligned}
\Delta(t) &= \max_{x,y} \frac{|\psi_y\,(P^t)\,\psi_x - \pi(x)|}{\pi(x)} \\
&= \max_{x,y} \frac{|\psi_y\,T^{-1/2}\,(I - \mathcal{L})^t\,T^{1/2}\,\psi_x - \pi(x)|}{\pi(x)} \\
&= \max_{x,y} |\psi_y T^{-1/2}(\sum_{i \neq 0}(1 - \lambda_i)^t I_i)T^{-1/2}\psi_x|\mathrm{vol}G \\
&\leq \max_{x,y} |\psi_y T^{-1/2}(H_t - I_0)T^{-1/2}\psi_x|\mathrm{vol}G \\
(10.25) \qquad &\leq \max_{x,y} |H_t(x,y)\frac{\mathrm{vol}\,G}{\sqrt{d_x d_y}} - 1|
\end{aligned}
$$

$$\square$$

Theorem 10.12 implies that an upper bound for the heat kernel can be used to bound the convergence rate for random walks. This and many other powerful consequences of the heat kernel will be discussed further in the next two chapters.

CHAPTER 11

Sobolev inequalities

11.1. The isoperimetric dimension of a graph

In spectral geometry, Sobolev inequalities are among the main tools for controlling eigenfunctions. In this chapter, we will derive the discrete versions of Sobolev inequalities. The proofs are self-contained and entirely graph-theoretic although many of the concepts and ideas can be traced back to the early work of Nash [254] on Riemannian manifolds. On one hand, graphs and Riemannian manifolds are quite different objects. Indeed, many of the theorems and proofs in differential geometry are very difficult to translate into similar ones for graphs (since there are no high-order derivatives on a graph). In fact, some of the statements of the theorems in the continuous cases are obviously not true in the discrete setting. On the other hand, there is a great deal of overlap between these two disparate areas both in concepts and methods. Some selected techniques in the continuous case can often be successfully carried out in the discrete domain. In particular, we will see that the methods of Sobolev inequalities for differential manifolds can be effectively utilized in dealing with general graphs. In Chapter 9, Harnack inequalities have been derived only for very special graphs (homogeneous graphs and their convex subgraphs). Here we will establish Sobolev inequalities for all graphs by using a key invariant, the isoperimetric dimension of a graph (also see [59]). The proof techniques in this chapter are quite similar to their parallels for the continuous case [49], [254].

We say that a graph G has *isoperimetric dimension* δ with *isoperimetric constant* c_δ if for every subset X of $V(G)$, the number of edges between X and the complement $\bar{X}$ of X, denoted by $|E(X, \bar{X})|$, satisfies

$$(11.1) \qquad |E(X, \bar{X})| \geq c_\delta (\text{vol } X)^{\frac{\delta - 1}{\delta}}$$

where vol $X \leq$ vol $\bar{X}$ and c_δ is a constant depending only on δ.

In other words, we can define the so-called *Sobolev constant* c_δ by

$$c_\delta = \inf_X \frac{|E(X, \bar{X})|}{(\text{vol } X)^{\frac{\delta - 1}{\delta}}}$$

where vol $X \leq$ vol $\bar{X}$.

We will prove two discrete versions of the Sobolev inequalities in Sections 11.2 and 11.3. We will show that for $\delta \geq 1$ and for any function $f : V(G) \to \mathbb{R}$,

(11.2)
$$\sum_{u \sim v} |f(u) - f(v)| \geq c_1 \min_m (\sum_v |f(v) - m|^{\frac{\delta}{\delta-1}} d_v)^{\frac{\delta-1}{\delta}}.$$

and for $\delta > 2$,

(11.3)
$$(\sum_{u \sim v} |f(u) - f(v)|^2)^{1/2} \geq c_2 \min_m (\sum_v |f(v) - m|^\gamma d_v)^{\frac{1}{\gamma}}$$

where $\gamma = \frac{2\delta}{\delta-2}$. Here the constants c_1 and c_2 depend only on δ. In fact we will show $c_1 \geq c_\delta \frac{\delta-1}{\delta}$ and $c_2 \geq \sqrt{c_\delta} \frac{\delta-1}{2\delta}$.

The Sobolev inequalities can be used to derive the following eigenvalue inequalities (see Section 11.4):

(11.4)
$$\sum_{i \neq 0} e^{-\lambda_i t} \leq c_3 \frac{\text{vol } G}{t^{\delta/2}}$$

and

(11.5)
$$\lambda_k \geq c_4 \left(\frac{k}{\text{vol } G} \right)^{\frac{2}{\delta}}$$

for suitable constants c_3 and c_4 which depend only on δ.

In a sense, a graph can be viewed as a discretization of a Riemannian manifold in $\mathbb{R}^n$ where n is roughly equal to δ. The eigenvalue bound in (11.5) is an analogue of Polya's conjecture [214] for Dirichlet eigenvalues of regular domains M in $\mathbb{R}^n$:

$$\lambda_k \geq \frac{2\pi}{w_n} \left(\frac{k}{\text{vol } M} \right)^{2/n}$$

where w_n is the volume of the unit disc in $\mathbb{R}^n$. Here, we consider Laplacians of general graphs and obtain eigenvalue estimates in terms of the isoperimetric dimension.

We remark that a closely related isoperimetric invariant [48] is the Cheeger constant h_G of a graph G (discussed in Chapter 2). In fact, the Cheeger constant can be viewed as a special case of the isoperimetric constant c_δ with $\delta = \infty$.

We note that the first Sobolev inequality (11.2) can be expressed as

$$\|\nabla f\|_1 \geq c_1 \|f\|_{\frac{\delta}{\delta-1}}$$

and the second one (11.3) as

$$\|\nabla f\|_2 \geq c_2 \|f\|_\gamma$$

for $\gamma = 2\delta/(\delta - 2)$. In general we can define the Sobolev constants $s_{p,q}$ of a graph, for $p, q > 0$, as follows:

$$s_{p,q} = \inf_f \sup_{c \in \mathbb{R}} \frac{\|\nabla f\|_p}{\|f - c\|_q}.$$

For example, the Cheeger constant h_G of a graph G is just $s_{1,1}$. The two Sobolev inequalities in (11.2) and (11.3) concern $s_{1,q}$ and $s_{2,q}$ for certain q depending on the isoperimetric dimension δ.

There is another concept related to isoperimetric invariants, the so-called "moderate growth rate" or "polynomial growth rate" condition [15] (also see [98]). For a vertex v in a graph G and an integer r, we define

$$N_r(v) = \{u \in V(G) \ : \ d(u,v) \leq r\}$$

where $d(u,v)$ denotes the distance between u and v. We say a graph G has polynomial growth rate of type (c, δ) if

$$\text{vol } N_r(v) \geq c \ r^\delta$$

for all vertices v and vol $N_r \leq$ vol $G/2$.

It can be shown that a graph with isoperimetric dimension δ and isoperimetric constant c_δ has polynomial growth rate of type (c', δ) for some c'. However, there are graphs which have polynomial growth rate of type (c, δ) but which do not have isoperimetric dimension δ. For example, consider the graph H which is formed by taking two copies of a graph having polynomial growth rate of type (c, δ) and joining them by an edge. It is easy to see that H has polynomial growth rate, but H cannot have isoperimetric dimension δ. In fact, this is an example which has polynomial growth rate, but its associated random walk is *not* rapidly mixing. On the other hand, graphs with bounded isoperimetric dimension have good eigenvalue bounds and are therefore rapidly mixing.

We remark that for symmetrical graphs, these two notions —polynomial growth rate and isoperimetric dimension— are basically the same. Using a result of Gromov [148] on the growth rate of a finitely generated group, Varopoulos [247] showed that a locally finite Cayley graph of an infinite group with a nilpotent subgroup of finite index has polynomial growth rate of type (c, δ) depending only on the structure of the group. An excellent survey on this topic can be found in [15].

11.2. An isoperimetric inequality

We will first prove the following.

THEOREM 11.1. *In a connected graph G with isoperimetric dimension $\delta > 1$ and isoperimetric constant c_δ, for an arbitrary function $f : V(G) \to \mathbb{R}$, let m denote the largest value such that*

$$\sum_{\substack{v \\ f(v)<m}} d_v \leq \sum_{\substack{u \\ f(u)\geq m}} d_u.$$

Then

$$\sum_{u \sim v} |f(u) - f(v)| \geq c_1 \Big(\sum_v |f(v) - m|^{\frac{\delta}{\delta-1}} d_v\Big)^{\frac{\delta-1}{\delta}}$$

where $c_1 = c_\delta \frac{\delta-1}{\delta}$.

Here we state two useful corollaries. The first one is an immediate consequence of Theorem 11.1 and the second one follows from the proof of Theorem 11.1.

COROLLARY 11.2. *In a connected graph G with isoperimetric dimension δ and isoperimetric constant c_δ, for an arbitrary function $f : V(G) \to \mathbb{R}$ we have*

$$\sum_{u \sim v} |f(u) - f(v)| \geq c_1 \min_m (\sum_v |f(v) - m|^{\frac{\delta}{\delta-1}} d_v)^{\frac{\delta-1}{\delta}}$$

where $c_1 = c_\delta \frac{\delta-1}{\delta}$.

COROLLARY 11.3. *In a connected graph G with isoperimetric dimension δ and isoperimetric constant c_δ, for a function $f : V(G) \to \mathbb{R}$ and a vertex w, define*

$$f_w(v) = \begin{cases} \max\{f(v), f(w)\} & \text{if } f(w) \geq m \\ \min\{f(v), f(w)\} & \text{if } f(w) < m \end{cases}$$

where m is as defined in Theorem 11.1. Then

$$\sum_{u \sim v} |f_w(u) - f_w(v)| + a_w |f(w) - m| \geq c_1 (\sum_v f_w(v)^{\frac{\delta}{\delta-1}})^{\frac{\delta-1}{\delta}}$$

where

$$c_1 = c_\delta \frac{\delta - 1}{\delta},$$

$$a_w = \begin{cases} |\{\{u, v\} \in E(G) : f(u) > f(w) \geq f(v)\}| & \text{if } f(w) \geq m \\ |\{\{u, v\} \in E(G) : f(u) \geq f(w) > f(v)\}| & \text{if } f(w) < m. \end{cases}$$

$$S_w = \begin{cases} \{v : f(v) \geq f(w)\} & \text{if } f(w) \geq m \\ \{u : f(u) \leq f(w)\} & \text{if } f(w) < m. \end{cases}$$

Roughly speaking, the proof of Theorem 11.1, although a bit lengthy, is just repeated applications of summation by parts (which, of course, is just the discrete version of integration by parts). Namely, for any x_i and y_i, for $i = 0, \cdots, n + 1$, the following simple equality holds:

$$(11.6) \qquad \sum_{i=0}^n (x_{i+1} - x_i) y_i = \sum_1^n x_i (y_{i-1} - y_i) + x_{n+1} y_n - x_0 y_0$$

and the last two terms vanish if $x_0 = y_n = 0$ or $y_0 = x_{n+1} = 0$. We will use (11.6) repeatedly in the following proof of Theorem 11.1:

PROOF. For a given function $f : V(G) \to \mathbb{R}$, we label the vertices so as to satisfy

$$f(v_1) \geq f(v_2) \geq \cdots \geq f(v_n).$$

We define $A_i = \{\{v_j, v_k\} \in E(G) : j \leq i < k\}$ and $a_i = |A_i|$. We use the notation $f(i) = f(v_i)$ and $d_i = d_{v_i}$. Define $S_i^- = \sum_{j < i} d_j$, $S_i^+ = \sum_{j \geq i} d_j$ and $S_i = \min\{S_i^-, S_i^+\}$. Clearly, $S_i = S_i^+$ for $f(i) \geq m$ and $S_i = S_i^-$ for $f(i) < m$. From the definition in (11.1), we have

$$a_i \geq c_\delta (\text{vol } S_i)^{\frac{\delta-1}{\delta}}.$$

We use the convention that $S_0 = S_{n+1} = 0$. Let $h(i) = h(v_i) = f(v_i) - m$, and suppose $f(i_0) = m$.

Then

$$\sum_{u \sim v} |h(u) - h(v)| \;=\; \sum_i a_i |h(i-1) - h(i)|$$

$$\geq\; c_\delta \sum_i S_i^{\frac{\delta-1}{\delta}} (h(i-1) - h(i))$$

$$\geq\; c_\delta \sum_{i < i_0} |h(i)|(S_{i+1}^{\frac{\delta-1}{\delta}} - S_i^{\frac{\delta-1}{\delta}})$$

$$+ c_\delta \sum_{i > i_0} |h(i)|(S_i^{\frac{\delta-1}{\delta}} - S_{i+1}^{\frac{\delta-1}{\delta}})$$

$$=\; W_1 + W_2$$

We consider W_1 first; W_2 can be dealt with in a similar way.

$$W_1 \;=\; c_\delta \sum_{i \leq i_0} h(i)((S_i + d_i)^{\frac{\delta-1}{\delta}} - S_i^{\frac{\delta-1}{\delta}})$$

$$\geq\; c_\delta \sum_{i \leq i_0} h(i) \frac{\delta-1}{\delta} \cdot \frac{d_i}{S_i^{\frac{1}{\delta}}}$$

$$\geq\; c_\delta \frac{\delta-1}{\delta} \sum_{i \leq i_0} \frac{h(i)^{\frac{\delta}{\delta-1}} d_i}{(h(i)^{\frac{\delta}{\delta-1}} S_i)^{1/\delta}}$$

$$\geq\; c_\delta \frac{\delta-1}{\delta} \frac{\displaystyle\sum_{i \leq i_0} |h(i)|^{\frac{\delta}{\delta-1}} d_i}{\left(\displaystyle\sum_{i \leq i_0} |h(i)|^{\frac{\delta}{\delta-1}} d_i \right)^{1/\delta}}$$

$$\geq\; c_\delta \frac{\delta-1}{\delta} \sum_{i \leq i_0} |h(i)|^{\frac{\delta}{\delta-1}} d_i)^{\frac{\delta-1}{\delta}}.$$

Combining this with the corresponding bound for W_2, we have

$$W_1 + W_2 \;=\; c_\delta \frac{\delta-1}{\delta} \left((\sum_{i \leq i_0} |h(i)|^{\frac{\delta}{\delta-1}} d_i)^{\frac{\delta-1}{\delta}} + (\sum_{i \geq i_0} |h(i)|^{\frac{\delta}{\delta-1}} d_i)^{\frac{\delta-1}{\delta}} \right)$$

$$\geq\; c_\delta \frac{\delta-1}{\delta} \left(\sum_i |h(i)|^{\frac{\delta}{\delta-1}} d_i \right)^{\frac{\delta-1}{\delta}}$$

and Theorem 11.1 is proved. $\square$

We remark that Corollary 11.3 follows from the above proof and the fact that for $f(w) \geq m$,

$$\sum_{u \sim v} |f_w(u) - f_w(v)| = \sum_{\substack{i \\ f(i) \geq f(w)}} a_i(h(i-1) - h(i))$$

$$\geq \sum_{\substack{i \\ h(i) \geq h(w)}} |h(i)|(a_i - a_{i-1}) - a_w |h(w)|.$$

11.3. Sobolev inequalities

The second Sobolev inequality (11.3) is more powerful than the first one (11.2). However, its proof in Theorem 11.4 is also more complicated than that of Theorem 11.1. Basically, the proof consists primarily of iterations of "summation by parts" using Theorem 11.1.

THEOREM 11.4. *For a graph G with isoperimetric dimension $\delta > 2$ and isoperimetric constant c_δ, any function $f : V(G) \to \mathbb{R}$ satisfies*

$$\left(\sum_{u \sim v} |f(u) - f(v)|^2\right)^{1/2} \geq c_2 \min_m \left(\sum_v |f(v) - m|^\gamma d_v\right)^{1/\gamma}$$

where $\gamma = \frac{2\delta}{\delta - 2}$ and $c_2 = \sqrt{c_\delta}\frac{\delta-1}{2\delta}$.

PROOF. We follow the notation in Theorem 11.1 where $h(i) = h(v_i) = f(v_i) - m$ and $f(i_0) = m$. Also, we denote $h_i(x) = h_{v_i}(x)$.

$$\sum_{u \sim v} |h(u) - h(v)|^2$$

$$= \sum_i (h(i-1) - h(i)) \sum_{\{j,k\} \in A_i} |h(j) - h(k)|$$

$$\geq \sum_{i \leq i_0} (h(i-1) - h(i)) \sum_{\{j,k\} \in A_i} |h_i(j) - h_i(k)|$$

$$+ \sum_{i \geq i_0} (h(i-1) - h(i)) \sum_{\{j,k\} \in A_i} |h_i(j) - h_i(k)|$$

$$= P_1 + P_2.$$

We will give a lower bound for the second part P_1 where without loss of generality we may assume that all $h(i)$'s are positive. The second part can be proved similarly.

$$P_1 \geq \sum_{i \leq i_0} h(i) \left(\sum_{\{j,k\} \in A_{i+1}} |h_{i+1}(j) - h_{i+1}(k)| - \sum_{\{j,k\} \in A_i} |h_i(j) - h_i(k)| \right)$$

$$\geq \sum_{i \leq i_0} h(i) \left(a_{i+1}(h(i) - h(i+1)) - \sum_{\{j,k\} \in A_i - A_{i+1}} |h_i(j) - h_i(k)| \right)$$

$$\geq \sum_{i \leq i_0} h(i) \, a_{i+1}(h(i) - h(i+1)) - P_1.$$

Now we use summation by parts again and we obtain

$$
\begin{aligned}
2P_1 &\geq \sum_{i \leq i_0} h(i) a_{i+1} (h(i) - h(i+1)) \\
&\geq \sum_{i \leq i_0} h(i) (a_{i+1} h(i) - a_i h(i-1)) \\
&\geq \sum_{i \leq i_0} h(i) ((a_{i+1} - a_i) h(i) - a_i (h(i-1) - h(i)) \\
&\geq \sum_{i \leq i_0} h(i) (a_{i+1} - a_i) h(i) - \sum_{i \leq i_0} h(i) (a_{i+1} (h(i) - h(i-1)) \\
&\geq \sum_{i \leq i_0} h(i) (a_{i+1} - a_i) h(i) - 2P_1
\end{aligned}
$$

We define

$$
T_i = \sum_{j \leq i} a_j (h(j-1) - h(j)) + a_i |h(i)|.
$$

From Cor. 11.3, we have

$$
\begin{aligned}
T_i &= \sum_{u \sim v} |f_{v_i}(u) - f_{v_i}(v)| + a_i (f(i) - m) \\
&\geq c_\delta \frac{\delta - 1}{\delta} \Big(\sum_{j < i} h(j)^{\frac{\delta}{\delta-1}} d_j \Big)^{\frac{\delta-1}{\delta}} \\
&:= c_\delta \frac{\delta - 1}{\delta} {T_i'}^{\frac{\delta-1}{\delta}}.
\end{aligned}
$$

It follows from the definition of T_i that

$$
T_i - T_{i-1} = (a_i - a_{i-1}) h(i-1).
$$

Hence,

$$
\begin{aligned}
P_1 &\geq \frac{1}{4} \sum_{i \leq i_0} h(i)(a_{i+1} - a_i) h(i) \\
&\geq \sum_{i \leq i_0} h(i)(T_{i+1} - T_i) \\
&\geq c_\delta \frac{\delta - 1}{4\delta} \sum_{i \leq i_0} h(i) T_i'^{\frac{\delta-1}{\delta}} \left(\left(1 + \frac{h(i)^{\frac{\delta}{\delta-1}} d_i}{T_i'}\right)^{\frac{\delta-1}{\delta}} - 1 \right) \\
&\geq c_\delta \frac{(\delta - 1)^2}{4\delta^2} \sum_{i \leq i_0} h(i) T_i'^{\frac{\delta-1}{\delta}} \frac{h(i)^{\frac{\delta}{\delta-1}} d_i}{T_i'} \\
&\geq c_\delta \frac{(\delta - 1)^2}{4\delta^2} \sum_{i \leq i_0} h(i) \frac{h(i)^{\frac{\delta}{\delta-1}} d_i}{T_i'^{\frac{1}{\delta}}} \\
&\geq c_\delta \frac{(\delta - 1)^2}{4\delta^2} \sum_{i \leq i_0} \frac{h(i)^{\frac{2\delta}{\delta-1}} d_i}{(\sum_{i < i_0} h(i)^{\frac{2\delta}{\delta-1}} d_i)^{\frac{1}{\delta}}} \\
&\geq c_\delta \frac{(\delta - 1)^2}{4\delta^2} \left(\sum_{i \leq i_0} h(i)^{\frac{2\delta}{\delta-1}} d_i \right)^{\frac{\delta-1}{\delta}} \\
&\geq c_\delta \frac{(\delta - 1)^2}{4\delta^2} \left(\sum_{i \leq i_0} h(i)^\gamma d_i \right)^{\frac{2}{\gamma}}.
\end{aligned}
$$

In a similar way we have

$$
P_2 \geq c_\delta \frac{(\delta - 1)^2}{4\delta^2} \left(\sum_{i \geq i_0} |h(i)|^\gamma d_i \right)^{\frac{2}{\gamma}}.
$$

Therefore,

$$
\left(\sum_{i \sim j} (h(i) - h(j))^2 \right)^{\frac{1}{2}} \geq \sqrt{c_\delta} \frac{\delta - 1}{2\delta} \left(\sum_i |h(i)|^\gamma d_i \right)^{\frac{1}{\gamma}}
$$

since $\gamma \geq 2$. $\square$

11.4. Eigenvalue bounds

In this section, we assume $\delta > 2$ since we will use the second Sobolev inequality to establish eigenvalue bounds.

Let H_t denote the heat kernel of G. From Section 10.2, the heat kernel satisfies

$$
H_t(x, y) = \sum_z H_s(x, z) H_{t-s}(z, y).
$$

In particular, we have

$$H_t(x,x) = \sum_y (H_{\frac{t}{2}}(x,y))^2.$$

We consider, for a fixed x,

$$\frac{\partial}{\partial t} H_t(x,x) = 2 \sum_y H_{\frac{t}{2}}(x,y) \frac{\partial}{\partial t} H_{\frac{t}{2}}(x,y)$$

$$= \sum_y H_{\frac{t}{2}}(x,y)(-\mathcal{L}H_{\frac{t}{2}}(y,x))$$

since the function $H_{\frac{t}{2}}(x,y)$, with y as the variable and x fixed, satisfies the heat equation in Lemma 10.3 (iii). Now we use Lemma 10.3 (iv) and get

$$\frac{\partial}{\partial t} H_t(x,x) = - \sum_{y \sim z} \left(\frac{H_{\frac{t}{2}}(y,x)}{\sqrt{d_y}} - \frac{H_{\frac{t}{2}}(z,x)}{\sqrt{d_z}} \right)^2$$

We apply Theorem 11.4 by again considering $H_{\frac{t}{2}}(y,x)$ as a function of y with fixed x. For $\gamma = \frac{2\delta}{\delta-2}$, we have

(11.7)
$$\frac{\partial}{\partial t} H_t(x,x) \leq -c_\delta \frac{(\delta-1)^2}{4\delta^2} \left(\sum_y \left(\frac{H_{\frac{t}{2}}(y,x)}{\sqrt{d_y}} - m \right)^\gamma d_y \right)^{2/\gamma}.$$

To proceed, we need an additional fact.

LEMMA 11.5.

$$\left(\sum_y \left(\frac{H_{\frac{2}{t}}(x,y)}{\sqrt{d_y}} - m \right)^\gamma d_y \right)^{\frac{1}{\gamma-1}} (3\sqrt{d_x})^{\frac{\gamma-2}{\gamma-1}}$$

$$\geq \sum_y \left(\frac{H_{\frac{t}{2}}(x,y)}{\sqrt{d_y}} - m \right)^2 d_y.$$

PROOF. We apply Hölder's inequality for $1 = \frac{1}{p} + \frac{1}{q}$,

$$\sum_i f_i g_i \leq \left(\sum_i f_i^p \right)^{1/p} \left(\sum_i g_i^q \right)^{1/q}$$

where we take $p = \gamma - 1, q = \frac{\gamma-1}{\gamma-2}$ and

$$f_y = \left| \frac{H_{\frac{t}{2}}(x,y)}{\sqrt{d_y}} - m \right|^{\frac{\gamma}{\gamma-1}},$$

$$g_y = \left| \frac{H_{\frac{t}{2}}(x,y)}{\sqrt{d_y}} - m \right|^{\frac{\gamma-2}{\gamma-1}}.$$

We then have

$$\left(\sum_y \left| \frac{H_{\frac{t}{2}}(x,y)}{\sqrt{d_y}} - m \right|^\gamma d_y \right)^{\frac{1}{\gamma-1}} \left(\sum_y \left| \frac{H_{\frac{t}{2}}(x,y)}{\sqrt{d_y}} - m \right| d_y \right)^{\frac{\gamma-2}{\gamma-1}}$$

$$\geq \sum_y \left(\frac{H_{\frac{t}{2}}(x,y)}{\sqrt{d_y}} - m \right)^2 d_y.$$

It remains to bound $\displaystyle\sum_y |\frac{H_{\frac{t}{2}}(x,y)}{\sqrt{d_y}} - m| d_y$ from above.

By using Lemma 10.3 (ii), we have

$$(11.8) \qquad \sum_y \left(\frac{H_{\frac{t}{2}}(x,y)}{\sqrt{d_y}} - m' \right) d_y \;=\; \sum_y H_{\frac{t}{2}}(x,y)\sqrt{d_y} - m' \mathrm{vol}\, G$$

$$=\; \sqrt{d_x} - m' \mathrm{vol}\, G$$

$$=\; 0$$

by choosing m' to be

$$m' = \frac{\sqrt{d_x}}{\mathrm{vol}\, G}.$$

Define $N_x^+ = \{y : \frac{H_{\frac{t}{2}}(y,x)}{\sqrt{d_y}} \geq m'\}$ and $N_x^- = \{y : \frac{H_{\frac{t}{2}}(y,x)}{\sqrt{d_y}} < m'\}$. Now

$$\sum_y |\frac{H_{\frac{t}{2}}(x,y)}{\sqrt{d_y}} - m'| d_y \;=\; \sum_{y \in N_x^+} \left(\frac{H_{\frac{t}{2}}(x,y)}{\sqrt{d_y}} - m' \right) d_y + \sum_{y \in N_x^-} \left(m' - \frac{H_{\frac{t}{2}}(x,y)}{\sqrt{d_y}} \right) d_y$$

$$=\; 2 \sum_{y \in N_x^-} \left(m' - \frac{H_{\frac{t}{2}}(x,y)}{\sqrt{d_y}} \right) d_y$$

$$\leq\; 2 \sum_{y \in N_x^-} m' d_y$$

$$=\; 2 \frac{\sqrt{d_x}}{\mathrm{vol}\, G} \cdot \sum_{y \in N_x^-} d_y$$

$$\leq\; 2\sqrt{d_x}.$$

Therefore,

$$\sum_y |\frac{H_{\frac{t}{2}}(x,y)}{\sqrt{d_y}} - m| d_y \;\leq\; \sum_y |\frac{H_{\frac{t}{2}}(x,y)}{\sqrt{d_y}} - m'| d_y + \sum_y |m' - m| d_y$$

$$\leq\; 2\sqrt{d_x} + \sum_y m' d_y$$

$$\leq\; 3\sqrt{d_x}$$

provided that $m' \geq m/2 > 0$. This, however, follows from the definition of m as in Theorem 11.1 and the fact that $H \geq 0$ and can be deduced as follows:

$$
\begin{aligned}
m' = \frac{\sqrt{d_x}}{\operatorname{vol} G} \;=\; & \frac{1}{\operatorname{vol} G} \sum_y H_{\frac{t}{2}}(x,y)\sqrt{d_y} \\[2mm]
=\; & \frac{1}{\operatorname{vol} G} \left(\sum_{y \in M_x^+} H_{\frac{t}{2}}(x,y)\sqrt{d_y} + \sum_{y \in M_x^-} H_{\frac{t}{2}}(x,y)\sqrt{d_y} \right) \\[2mm]
\geq\; & \frac{1}{\operatorname{vol} G} \sum_{y \in M_x^+} H_{\frac{t}{2}}(x,y)\sqrt{d_y} \\[2mm]
\geq\; & \frac{m \operatorname{vol} M_x^+}{\operatorname{vol} G} \\[2mm]
\geq\; & \frac{m}{2}
\end{aligned}
$$

where $M_x^+ = \{y : \frac{H_{\frac{t}{2}}(y,x)}{\sqrt{d_y}} \geq m\}$ and $M_x^- = \{y : \frac{H_{\frac{t}{2}}(y,x)}{\sqrt{d_y}} < m\}$. Recall from the definition of m, we have $\operatorname{vol} M_x^+ \geq \operatorname{vol} G/2$. This completes the proof of Lemma 11.5. $\qquad\qquad\square$

We note that

$$
\sum_y \left(\frac{H_{\frac{t}{2}}(x,y)}{\sqrt{d_y}} - m \right)^2 d_y \geq \sum_y \left(\frac{H_{\frac{t}{2}}(x,y)}{\sqrt{d_y}} - m' \right)^2 d_y
$$

because of (11.8).

We now return to inequality (11.7). Using Lemma 11.5, we obtain

$$
\begin{aligned}
\frac{\partial}{\partial t} & H_t(x,x) \\[2mm]
\leq\; & -c_5 \left(\sum_y \left(\frac{H_{\frac{t}{2}}(x,y)}{\sqrt{d_y}} - m \right)^{\gamma} d_y \right)^{2/\gamma} \\[3mm]
\leq\; & -c_5 \left(\sum_y \left(\frac{H_{\frac{t}{2}}(x,y)}{\sqrt{d_y}} - m \right)^{2} d_y \right)^{\frac{2(\gamma-1)}{\gamma}} (3\sqrt{d_x})^{\frac{-2(\gamma-2)}{\gamma}} \\[3mm]
\leq\; & -c_5 \left(\sum_y \left(\frac{H_{\frac{t}{2}}(x,y)}{\sqrt{d_y}} - m' \right)^{2} d_y \right)^{\frac{2(\gamma-1)}{\gamma}} (3\sqrt{d_x})^{\frac{-2(\gamma-2)}{\gamma}} \\[3mm]
\leq\; & -c_5 \left(\sum_y (H_{\frac{t}{2}}(x,y)^2 - 2m' H_{\frac{t}{2}}(x,y)\sqrt{d_y} + m'^2 d_y) \right)^{\frac{2(\gamma-1)}{\gamma}} (3\sqrt{d_x})^{-\frac{2(\gamma-2)}{\gamma}} \\[3mm]
\leq\; & -c_5 (H_t(x,x) - 2m'\sqrt{d_x} + m'^2 \operatorname{vol} G)^{2(\frac{\gamma-1}{\gamma})} (3\sqrt{d_x})^{-\frac{2(\gamma-2)}{\gamma}} \\[3mm]
\leq\; & -c_5 (H_t(x,x) - \frac{d_x}{\operatorname{vol} G})^{2\frac{(\gamma-1)}{\gamma}} (3\sqrt{d_x})^{-\frac{2(\gamma-2)}{\gamma}}
\end{aligned}
$$

where $c_5 = (c_\delta \frac{(\delta-1)^2}{2\delta^2})^{2(\gamma-1)/\gamma}$. We note that $1 - 2(\frac{\gamma-1}{\gamma}) = -\frac{2}{\delta}$. We consider

$$\frac{\partial}{\partial t}(H_t(x,x) - \frac{d_x}{\text{vol } G})^{-\frac{2}{\delta}}$$

$$\frac{\partial}{\partial t}(H_t(x,x) - \frac{d_x}{\text{vol } G})^{1-2\frac{(\gamma-1)}{\gamma}}$$

$$= -\frac{2}{\delta}(H_t(x,x) - \frac{d_x}{\text{vol } G})^{-2(\frac{\gamma-1}{\gamma})}\frac{\partial}{\partial t}H_t(x,x)$$

$$\geq \frac{2c_5}{\delta}(H_t(x,x) - \frac{d_x}{\text{vol } G})^{-2(\frac{\gamma-1}{\gamma})+2(\frac{\gamma-1}{\gamma})}(3\sqrt{d_x})^{-2(\frac{\gamma-2}{\gamma})}$$

$$\geq \frac{2c_5}{\delta}(3\sqrt{d_x})^{-2(\frac{\gamma-2}{\gamma})}$$

using the preceding inequality for $\frac{\partial}{\partial t}H_t(x,x)$. Therefore, we obtain

$$(H_t(x,x) - \frac{d_x}{\text{vol } G})^{-\frac{2}{\delta}} \geq c_6(3\sqrt{d_x})^{-2(\frac{\gamma-2}{\gamma})}t + (H_0(x,x) - \frac{d_x}{\text{vol } G})^{-\frac{2}{\delta}}$$

$$\geq c_6(3\sqrt{d_x})^{-2(\frac{\gamma-2}{\gamma})}t + (1 - \frac{d_x}{\text{vol } G})^{-\frac{2}{\delta}}$$

$$\geq c_6(3\sqrt{d_x})^{-2(\frac{\gamma-2}{\gamma})}t$$

where $c_6 = 2c_5/\delta$. This is equivalent to

$$H_t(x,x) - \frac{d_x}{\text{vol } G} \leq \frac{c_3 d_x}{t^{\frac{\delta}{2}}}$$

where $c_3 = 9c_6^{\delta/2}$. Hence,

$$\sum_x H_t(x,x) - 1 \leq \frac{c_3 \text{vol } G}{t^{\frac{\delta}{2}}}.$$

Since

$$\sum_x H_t(x,x) = \sum_x(\sum_i e^{-\lambda_i t}\phi_i^2(x)) = \sum_i e^{-\lambda_i t},$$

we have proved the following:

THEOREM 11.6.

(11.9) $$\sum_{i \neq 0} e^{-\lambda_i t} \leq \frac{c_3 \text{vol } G}{t^{\frac{\delta}{2}}}$$

where c_3 is a constant depending on δ.

From Theorem 11.6, we derive bounds for eigenvalues.

THEOREM 11.7. The k-th eigenvalue λ_k of $\mathcal{L}$ satisfies

$$\lambda_k \geq c_4(\frac{k}{\text{vol } G})^{2/\delta}$$

where c_4 is a constant depending on δ.

PROOF. From (11.6) we have

$$ke^{-\lambda_k t} \leq \frac{c_3 \text{vol } G}{t^{\frac{\delta}{2}}}$$

for all $t > 0$. Therefore

$$
\begin{aligned}
k & \leq c_3 \text{vol } G \cdot \inf_t \frac{e^{\lambda_k t}}{t^{\frac{\delta}{2}}} \\
& = c_3 \text{vol } G \cdot (\frac{2\lambda_k e}{\delta})^{\frac{\delta}{2}}
\end{aligned}
$$

since the function $\frac{e^{\lambda_k t}}{t^{\frac{\delta}{2}}}$ is minimized when $t = \frac{\delta}{2\lambda_k}$. This implies

$$
\begin{aligned}
\lambda_k & \geq \frac{\delta}{2e}(\frac{k}{c_3 \text{vol } G})^{\frac{2}{\delta}} \\
& = c_4(\frac{k}{\text{vol } G})^{\frac{2}{\delta}}
\end{aligned}
$$

where $c_4 = \frac{\delta}{2e}c_3^{2/\delta}$. $\qquad\qquad\square$

We remark that the constants c_3 and c_4 can be explicitly computed from the proofs but are somewhat messy expressions. Here we derive these inequalities with no intention of carefully optimizing the constants.

11.5. Generalizations to weighted graphs and subgraphs

In this section, we consider weighted undirected graphs G with edge weights $w_{u,v} = w_{v,u}$ for vertices u, v of G. We can define the isoperimetric dimension of G in a similar way: We say that G has isoperimetric dimension δ and isoperimetric constant c_δ if

$$
\sum_{u \in X} \sum_{v \in \bar{X}} w_{u,v} \geq c_\delta (\text{vol } X)^{1-\frac{1}{\delta}}
$$

for all $X \subseteq V(G)$ with $\text{vol } X \leq \text{vol } \bar{X}$ where $d_v = \sum_u w_{u,v}$, and $\text{vol } X = \sum_{v \in X} d_v$.

The results in previous sections can be generalized for the Laplacians of weighted undirected graphs with boundary conditions. We will state these facts but omit the proofs which follow along the lines in the previous sections.

THEOREM 11.8. *Let G be a weighted undirected graph with edge weights $w_{u,v}$ for vertices u, v of G, and suppose G has isoperimetric dimension δ and isoperimetric constant c_δ. Let S denote a subset of vertices of G. Then, any function $f : S \cup \delta S \to \mathbb{R}$ with either Dirichlet or Neumann boundary conditions satisfies*

$$
\sum_{\{u,v\} \in S^*} |f(u) - f(v)|w_{u,v} \geq c_1 \min_m (\sum_{v \in S} |f(v) - m|^{\frac{\delta}{\delta-1}} d_v)^{\frac{\delta-1}{\delta}}
$$

where S^* consists of all edges with at least one endpoint in S, and $c_1 = c_\delta \frac{\delta-1}{2\delta}$.

THEOREM 11.9. *Let G be a weighted undirected graph with edge weights $w_{u,v}$ for vertices u, v of G, and suppose G has isoperimetric dimension $\delta > 2$ and isoperimetric constant c_δ. Then any function $f : V(G) \to \mathbb{R}$ with either Dirichlet or Neumann boundary conditions satisfies*

$$
(\sum_{\{u,v\} \in S^*} |f(u) - f(v)|^2 w_{u,v})^{1/2} \geq c_2 \min_m (\sum_{v \in S} |f(v) - m|^\gamma d_v)^{1/\gamma}
$$

where $\gamma = \frac{2\delta}{\delta-2}$ *and* $c_2 = \sqrt{c_\delta}\frac{\delta-1}{2\delta}$.

THEOREM 11.10. *For a weighted undirected graph G and an induced subgraph S, the Dirichlet or Neumann eigenvalues of S satisfy*

$$\sum_{i\neq 0} e^{-\lambda_i t} \leq \frac{c_3 \text{vol } S}{t^{\delta/2}}$$

where c_3 is a constant depending only on δ.

THEOREM 11.11. *Suppose a weighted undirected graph G has isoperimetric dimension δ and isoperimetric constant c_δ. Let S denote an induced subgraph. The k-th Dirichlet or Neumann eigenvalue of S satisfies*

$$\lambda_k \geq c_4 \left(\frac{k}{\text{vol } G}\right)^{2/\delta}$$

where c_4 is a constant depending only on δ.

Advanced techniques for random walks on graphs

12.1. Several approaches for bounding convergence

Suppose in a graph G with edge weights $w_{x,y}$, a random walk $(x_0, x_1, \cdots, x_s)$ has transition probability

$$P(u, v) = Prob(x_{i+1} = v \mid x_i = u) = \frac{w_{u,v}}{d_u}$$

where $d_u = \sum_v w_{u,v}$. The number of steps required for a random walk to converge to the stationary distribution π, where

$$\pi(x) = \frac{d_x}{\text{vol } G}, \quad \text{vol } G = \sum_x d_x,$$

is closely related to the eigenvalues of the Laplacian (see Sections 1.5, 10.5). In particular, suppose we define

$$(12.1) \qquad \lambda = \begin{cases} \lambda_1 & \text{if } 1 - \lambda_1 \geq \lambda_{n-1} - 1 \\ 2 - \lambda_{n-1} & \text{otherwise.} \end{cases}$$

Then the relative pointwise distance of P to π after s steps satisfies

$$\Delta(s) = \max_{x,y} \frac{|P^s(y, x) - \pi(x)|}{\pi(x)} \leq e^{-c}$$

if

$$(12.2) \qquad s \geq \frac{1}{\lambda} \log \frac{\text{vol } G}{\min_x d_x} + \frac{c}{\lambda}.$$

We remark that this can be slightly improved by using the lazy walk as described in Theorem 1.16 and Corollary 1.17. Therefore, the λ in (12.2) can be taken to be

$$(12.3) \qquad \lambda = \begin{cases} \lambda_1 & \text{if } 1 - \lambda_1 \geq \lambda_{n-1} - 1 \\ \dfrac{2\lambda_1}{\lambda_{n-1} + \lambda_1} & \text{otherwise} \end{cases}$$

Here λ_i's are eigenvalues of the Laplacian of the graph G (see definitions in Chapter 1).

In this chapter, we will describe several methods for reducing the factor $\log \frac{\text{vol } G}{\min_x d_x}$ in (12.2) by using the *log-Sobolev constant*.

Suppose $G = (V, E)$ is a graph with edge weights $w_{x,y}$ for $x, y \in V$. The log-Sobolev constant α of G is the largest constant satisfying the following *log-Sobolev inequality* for any nontrivial function $f : V \to \mathbb{R}$:

$$\sum_{\{x,y\}\in E} (f(x) - f(y))^2 w_{x,y} \geq \alpha \sum_{x\in V} f^2(x) d_x \log \frac{f^2(x)\mathrm{vol}\ G}{\sum_{z\in V} f^2(z) d_z}$$

In other words, α can be expressed as follows:

$$(12.4) \qquad \alpha_G = \alpha = \inf_{f\neq 0} \frac{\sum_{\{x,y\}\in E} (f(x) - f(y))^2 w_{x,y}}{\sum_{x\in V} f^2(x) d_x \log \dfrac{f^2(x)\mathrm{vol}\ G}{\sum_{z\in V} f^2(z) d_z}}$$

where f ranges over all nontrivial functions $f : V \to \mathbb{R}$.

Logarithmic Sobolev inequalities first arose in the analysis of infinite dimensional elliptic differential operators. Many developments and applications can be found in the survey papers [**16, 149, 150, 234**]. Recently, Diaconis and Saloff-Coste [**101**] used a discrete version of the logarithmic Sobolev inequality for improving convergence bounds for Markov chains. In Section 12.2, we will improve (12.2) to:

$$\Delta(t) \leq e^{2-c}$$

if

$$(12.5) \qquad t \geq \frac{1}{2\alpha} \log\log \frac{\mathrm{vol}G}{\min_x d_x} + \frac{c}{\lambda}.$$

Also we will show

$$\Delta_{TV}(t) \leq e^{1-c}$$

if

$$(12.6) \qquad t \geq \frac{1}{4\alpha} \log\log \frac{\mathrm{vol}G}{\min_x d_x} + \frac{c}{\lambda}.$$

This is a slight improvement of [**101**] in which Diaconis and Saloff-Coste proved that

$$\Delta_{TV}(t) \leq e^{1-c}$$

if

$$t \geq \frac{1}{2\alpha} \log\log n + \frac{c}{\lambda}$$

for regular graphs. So, lower bounds for log-Sobolev constants can be used to improve convergence bounds for random walks on graphs and certain induced subgraphs. However, the problem of lower bounding log-Sobolev constants seems to be harder than finding eigenvalue bounds (even for special graphs). One relatively easy method is to use comparison theorems that will be discussed in Section 12.2.

A direct and powerful approach for estimating the log-Sobolev constant is by using the *logarithmic Harnack inequalities*. In Section 12.4, we will establish a logarithmic Harnack inequality which has a similar flavor to the Harnack inequality (discussed in Chapter 9). This provides an effective method for controlling the behavior of the function achieving the log-Sobolev constant. However, the proofs here hold only for homogeneous graphs and their special subgraphs.

Another approach for bounding the convergence rate for random walks in a graph is by using Sobolev inequalities and the isoperimetric dimension of a graph (defined in Section 11.1). This will be discussed in Section 12.5. As a consequence, for graphs with bounded isoperimetric dimension, their random walks have particularly nice bounds for convergence.

The definition for the log-Sobolev constant for a graph G can be easily generalized to induced subgraphs with boundary conditions. Let S denote a subset of vertices in a graph G and let S^* denote the set of edges with at least one endpoint in S. Let δS denote the vertex boundary of S. The log-Sobolev constant $\alpha_S^{(D)}$ for the induced subgraph S with Dirichlet boundary condition can be defined as follows:

$$(12.7) \qquad \alpha_S^{(D)} = \inf_f \frac{\displaystyle\sum_{\{x,y\}\in S^*} (f(x) - f(y))^2 w_{x,y}}{\displaystyle\sum_{x\in V} f^2(x)d_x \log \frac{f^2(x)\operatorname{vol} S}{\displaystyle\sum_{z\in S} f^2(z)d_z}}$$

where f ranges over all nontrivial functions $f : S \cup \delta S \to \mathbb{R}$ satisfying $f(y) = 0$ for $y \in \delta S$.

Also, the log-Sobolev constant α_S for the induced subgraph S with Neumann boundary condition can be defined as follows:

$$(12.8) \qquad \alpha_S = \inf_{f\neq c} \frac{\displaystyle\sum_{\{x,y\}\in S^*} (f(x) - f(y))^2 w_{x,y}}{\displaystyle\sum_{x\in V} f^2(x)d_x \log \frac{f^2(x)\operatorname{vol} S}{\displaystyle\sum_{z\in S} f^2(z)d_z}}$$

where f ranges over all non-constant functions $f : S \cup \delta S \to \mathbb{R}$. Many methods for bounding log-Sobolev constants for graphs can be extended to the log-Sobolev constant for certain subgraphs as well.

A relatively easy upper bound for the log-Sobolev constant is half of the first eigenvalue λ_1 of the Laplacian $\mathcal{L}$ of G.

LEMMA 12.1. *For a graph G (with no boundary or with Dirichlet or Neumann boundary conditions), we have*

$$(12.9) \qquad \alpha \leq \frac{\lambda_1}{2}.$$

PROOF. Let f denote a harmonic eigenfunction associated with eigenvalue λ_1. We consider $f' = 1 + \epsilon f$ for some small $\epsilon > 0$. Since $\sum_x f(x)d_x = 0$, we have

$$\sum_x f'^2(x)d_x = \operatorname{vol} G \; + \; \epsilon^2 \sum_x f^2(x)d_x.$$

From the definition of α (see (12.4)), we have

$$\alpha \leq \frac{\displaystyle\sum_{\{x,y\}\in E} (f'(x) - f'(y))^2 w_{x,y}}{\displaystyle\sum_{x\in V} f'^2(x) d_x \log \frac{f'^2(x)\text{vol } G}{\displaystyle\sum_{x\in V} f'^2(x)d_x}} = \frac{I}{II}$$

We consider

$$I = \sum_{\{x,y\}\in E} (f'(x) - f'(y))^2 w_{x,y} = \epsilon^2 \sum_{\{x,y\}\in E} (f(x) - f(y))^2 w_{x,y}.$$

Also, by repeatedly using $\displaystyle\sum_x f(x)d_x = 0$ and $\log(1+x) = x - x^2/2 + O(x^3)$, we have

$$II = 2\epsilon^2 \sum_x f^2(x) d_x + O(\epsilon^3).$$

By taking ϵ close to 0, we obtain

$$\alpha \leq \frac{\lambda_1}{2}.$$

$\square$

There are many examples which show that equality in (12.9) does not hold [**98**]. Suppose (12.9) holds or even that the log-Sobolev constant is within a constant factor of λ_1. Then the random walk on G converges to its stationary distribution in order $O(\lambda_1^{-1} \log\log(\text{vol } G/\max \text{ degree}))$. For a regular graph on n vertices, the number of steps required is of order $\lambda_1^{-1} \log\log n$. For example, for the n-cube Q_n, it can be shown that its log-Sobolev constant is equal to $\lambda_1/2 = 1/n$. Therefore the lazy random walk on the cube Q_n (with 2^n vertices) converges in time order $n \log n$. We remark that just as the eigenvalues of a weighted cartesian product of graphs can be determined (see Section 2.6 and Theorem 2.13), i.e.,

$$\lambda_{G \otimes H} = \frac{1}{2} \min(\lambda_G, \lambda_H),$$

the log-Sobolev constants also also have the same property:

$$\alpha_{G \otimes H} = \frac{1}{2} \min(\alpha_G, \alpha_H).$$

This can be used to determine the log-Sobolev constant for Q_n and will be left as an exercise.

12.2. Logarithmic Sobolev inequalities

Let G denote a weighted graph on n vertices. For a function $f : V(G) \to \mathbb{R}$, we may view f as a column vector, or a $1 \times n$ matrix as a row vector. The stationary distribution $\pi(x) = d_x/\text{vol } G$ will be regarded as a column or row vector. Let π denote the diagonal matrix with value $\pi(x) = d_x/\text{vol } G$ as its (x, x)-entry.

For a function $f : V(G) \rightarrow \mathbb{R}$, we define the $(\pi; p)$-norm of f, denoted by $_\pi \|f\|_p$, to be

$$_\pi \|f\|_p = \left(\sum_{x \in V(G)} f^p(x) \pi(x) \right)^{1/p}.$$

In particular,

$$_\pi \|f\|_2 = \left(\sum_x f^2(x) \pi(x) \right)^{1/2} = \|\pi^{1/2} f\|_2.$$

The main proof for (12.5) consists of two parts. In the first part (Theorem 12.2), we will see that the inequality (12.10) relating the p-norm to the 2-norm for certain p, implies the improved convergence bound for random walks. The second part (Theorem 12.4) states that the inequality (12.10) can be derived from the log-Sobolev inequality.

THEOREM 12.2. *Suppose that in a weighted graph G, its heat kernel H_s satisfies*

(12.10) $$_\pi \|f \, \pi^{1/2} \, H_s \, \pi^{-1/2}\|_p \leq \ _\pi \|f\|_2$$

for all $f : V(G) \rightarrow \mathbb{R}$, and $p = e^{\beta s}$ for some positive value β. Then the random walk on G satisfies

$$\Delta(t) \leq e^{2-c}$$

if

$$t \geq \frac{2}{\beta} \log\log \frac{\mathrm{vol} G}{\min_x d_x} + \frac{c}{\lambda}$$

where λ is as defined in (12.1) (or (12.3) for the modified walk defined in (1.15)).

PROOF. We define q by

$$\frac{1}{p} + \frac{1}{q} = 1.$$

For a vertex x of G, let ψ_x denote the characteristic function satisfying $\psi_x(y) = 1$ if $x = y$, and 0 otherwise. For a function $f : V \rightarrow \mathbb{R}$, we consider

(12.11) $$|\psi_x \pi^{-1/2} H_s \pi^{1/2} f|$$
$$= |\left(\psi_x \pi^{-1+1/q} \right) (\pi^{1/p-1/2} H_s \pi^{1/2} f)|$$
$$\leq \left(\sum_y (\psi_x \, \pi^{-1}(y))^q \pi(y) \right)^{1/q} \left(\sum_y (f \pi^{1/2} H_s \, \pi^{-1/2}(y))^p \pi(y) \right)^{1/p}$$
$$= \ _\pi \|\psi_x \pi^{-1}\|_q \ _\pi \|f \pi^{1/2} H_s \pi^{-1/2}\|_p$$

by using Hölder's inequality.

We consider

$$
\begin{aligned}
_\pi\|\psi_x \pi^{-1}\|_q &= \left(\sum_y (\psi_x\,\pi^{-1}(y))^q \pi(y)\right)^{1/q} \\
&= (\pi(x)^{1-q})^{1/q} \\
&= \pi(x)^{-1/p} \\
&\leq \left(\frac{\mathrm{vol}G}{\min_x d_x}\right)^{1/p}.
\end{aligned}
$$

Using the hypothesis that $p = e^{\beta s}$ and choosing s as

$$
s = \frac{1}{\beta}\log\log\frac{\mathrm{vol}G}{\min_x d_x},
$$

we have

$$
\left(\frac{\mathrm{vol}G}{\min_x d_x}\right)^{1/p} = e^{e^{\log\log\frac{\mathrm{vol}G}{\min_x d_x} - \beta s}} = e.
$$

From (12.10) and (12.11), we have, for any f,

$$
\begin{aligned}
|\psi_x \pi^{-1/2} H_s \pi^{1/2} f| &\leq e\,_\pi\|f\pi^{1/2}H_s\pi^{-1/2}\|_p \\
&\leq e\,_\pi\|f\|_2.
\end{aligned}
$$

In particular, for the heat kernel and the projection I_0 into the zeroth eigenfunction, we have

$$
\begin{aligned}
|\psi_x \pi^{-1/2}(H_{s+r} - I_0)\pi^{1/2} f| &\leq |\psi_x \pi^{-1/2} H_s(H_r - I_0)\pi^{1/2} f| \\
&\leq e\,_\pi\|\pi^{-1/2}(H_r - I_0)\pi^{1/2} f\|_2 \\
&\leq e\|(H_r - I_0)\pi^{1/2} f\|_2 \\
&\leq e\|(H_r - I_0)\|_2\,\|\pi^{1/2}f\|_2 \\
&\leq e^{1-\lambda r}\,_\pi\|f\|_2.
\end{aligned}
$$

This is equivalent to

$$
|\psi_x \pi^{-1/2}(H_{s+r} - I_0)g| \leq e^{1-\lambda r}\|g\|_2
$$

for all g. This implies

(12.12) $$\|\psi_x \pi^{-1/2}(H_{s+r} - I_0)\|_2 \leq e^{1-\lambda r}.$$

Therefore, the random walk on G converges to the stationary distribution under relative pairwise distance as follows (see (10.23)):

$$
\begin{aligned}
\Delta(2s + 2r) &\leq \max_{x,y} |\psi_x \pi^{-1/2}(H_{2s+2r} - I_0)\pi^{-1/2}\psi_y| \\
&\leq \max_{x,y}\|\psi_x \pi^{-1/2}(H_{s+r} - I_0)\|_2 \cdot \|\psi_y \pi^{-1/2}(H_{s+r} - I_0)\|_2 \\
&\leq e^{2-2\lambda r}
\end{aligned}
$$

by using (12.12) and the Cauchy-Schwarz inequality. Now, we take $r = \frac{c}{2\lambda}$, $t = 2s + 2r$, and the proof is complete. $\square$

We can also obtain a similar statement for the convergence bound under the total variation distance.

THEOREM 12.3. *Suppose that in a weighted graph G, its heat kernel H_s satisfies*

$$_\pi\|f \, \pi^{1/2} \, H_s \, \pi^{-1/2}\|_p \; \le \; _\pi\|f\|_2$$

for all $f : V(G) \to \mathbb{R}$, and $p = e^{\beta s}$ for some positive value β. Then the random walk on G satisfies

$$\Delta_{TV}(t) \le \frac{1}{2} e^{1-c}$$

if

$$t \ge \frac{1}{\beta} \log\log \frac{\mathrm{vol}G}{\min_x d_x} + \frac{c}{\lambda}$$

where λ is as defined in (12.1) (or (12.3) for the modified walk as in (1.15)).

PROOF. We follow the notation in Theorem 12.2.

$$
\begin{aligned}
\Delta_{TV} &= \frac{1}{2} \max_x \sum_y |\psi_x P^{s+r}(y) - \pi(y)| \\
&\le \frac{1}{2} \max_x \sum_y |\psi_x \pi^{-1/2}(H_{s+r} - I_0)\pi^{1/2}(y)| \\
&\le \frac{1}{2} \max_x \sum_y e^{1-\lambda r}\pi(y) \\
&\le \frac{1}{2} e^{1-\lambda r}
\end{aligned}
$$

by using (12.12).   □

Now we proceed to show that the log-Sobolev constant can be used to determine β in the above theorems. This proof is very similar to the continuous case ([**149**]; also see [**101**]).

THEOREM 12.4. *In a graph G with log-Sobolev constant α, its heat kernel H_t satisfies*

$$_\pi\|f \, \pi^{1/2} \, H_t \, \pi^{-1/2}\|_p \; \le \; _\pi\|f\|_2$$

for any $t > 0$, $p = e^{4\alpha t} + 1$, and for any $f : V(G) \to \mathbb{R}$.

PROOF. From the definition of α, we have

$$\sum_{x \sim y}(f(x) - f(y))^2 w(x,y) \ge \alpha \sum_x f^2(x)d_x \log \frac{f^2(x)^2}{\sum_z f^2(z)\pi(z)}$$

for any nontrivial function f. In particular, we can replace f by $f^{p/2}$ and we obtain

$$(12.13) \qquad \sum_{x \sim y}(f^{p/2}(x) - f^{p/2}(y))^2 w(x,y) \ge \alpha \sum_x f^p(x)d_x \log \frac{f(x)^p}{\sum_z f^p(z)\pi(z)}.$$

Now we need the following inequality which is not hard to prove:

$$(12.14) \qquad 4(p-1)(a^{p/2} - b^{p/2})^2 \le p^2(a-b)(a^{p-1} - b^{p-1}).$$

for all $a, b \geq 0$ and $p \geq 1$. From (12.13) and (12.14), we have

$$\alpha \sum_x f^p(x)\pi(x) \log \frac{f(x)^p}{\sum_z f^p(z)\pi(z)}$$

$$\leq \sum_{x \sim y}(f^{p/2}(x) - f^{p/2}(y))^2 w(x,y)$$

$$(12.15) \qquad \leq \frac{p^2}{4(p-1)} \sum_{x \sim y}(f^{p-1}(x) - f^{p-1}(y))(f(x) - f(y))w_{x,y}$$

We now replace f by $g = f\pi^{1/2}H_t\pi^{-1/2}$ in the above inequality and define p as a function of t:

$$p = p(t) = 1 + e^{4\alpha t}.$$

Note that $p' = p'(t) = 4\alpha(p-1)$. From (12.15), we have

(12.16)

$$\frac{p'}{p^2} \sum_x g^p(x)\pi(x) \log \frac{|g(x)|^p}{\sum_z g^p(z)\pi(z)} - \sum_{x \sim y}(g^{p-1}(x) - g^{p-1}(y))(g(x) - g(y))w_{x,y} \leq 0.$$

Now we define

$$F(t) = {}_\pi\|g\|_p.$$

Clearly, $F(0) = {}_\pi\|f\|_2$. If we can show that the derivative $F'(t) \leq 0$, then we have ${}_\pi\|g\|_p = F(t) \leq F(0) = {}_\pi\|f\|_2$ as desired. It remains to show $F'(t) \leq 0$. Since

$$F(t) = \left(\sum_x (f\pi^{1/2}H_t\pi^{-1/2}(x))^p\pi(x)\right)^{1/p} = (G(t))^{1/p},$$

we have

$$(12.17) \qquad F'(t) = \left(-\frac{p'}{p^2}\log G(t) + \frac{G'(t)}{pG(t)}\right)F(t).$$

We note that

$$G'(t) = p\sum_x g^{p-1}\pi(x)(f\pi^{1/2}\frac{d}{dt}H_t\pi^{-1/2}(x)) + p'\sum_x g^p(x)\pi(x)\log g(x)$$

$$= I + II$$

We consider the above sum of I as a product of matrices (where A^* denotes the transpose of A):

$$I = p\, g^{p-1}\pi^{1/2}\frac{d}{dt}H_t\pi^{1/2}f^*$$

$$= -p\, g^{p-1}\pi^{1/2}\mathcal{L}H_t\pi^{1/2}f^*$$

$$= -p\, g^{p-1}\pi^{1/2}\mathcal{L}\pi^{1/2}g^*$$

$$= -\frac{p}{\text{vol } G}\sum_{x \sim y}(g^{p-1}(x) - g^{p-1}(y))(g(x) - g(y))w_{x,y}$$

by using the heat equation in the weighted version of Lemma 10.3. Substituting into (12.17), we obtain

$$
\begin{aligned}
F'(t) \;=\; & \frac{p'}{p^2}\Big(\sum_x g^p(x)\pi(x)\log g^p(x) - \log G(t)\Big) \\[2mm]
& - \frac{1}{\operatorname{vol} G}\sum_{x\sim y}(g^{p-1}(x) - g^{p-1}(y))(g(x) - g(y)) \\[2mm]
\;=\; & \frac{1}{\operatorname{vol} G}\Big(\frac{p'}{p^2}\sum_x\sum_x g^p(x)d_x\log\frac{g^p(x)}{\log G(t)} - \\[2mm]
& \sum_{x\sim y}(g^{p-1}(x) - g^{p-1}(y))(g(x) - g(y))w_{x,y}\Big) \\[2mm]
\;\le\; & 0
\end{aligned}
$$

by using (12.16). This completes the proof of Theorem 12.4.   □

Therefore we have

THEOREM 12.5. *In a weighted graph G with log-Sobolev constant α, we have* $\Delta(t)\le e^{2-c}$ *if*

$$
t \ge \frac{1}{2\alpha}\log\log\frac{\operatorname{vol} G}{\min_x d_x} + \frac{c}{\lambda}.
$$

THEOREM 12.6. *In a weighted graph G with log-Sobolev constant α, we have* $\Delta_{TV}(t)\le e^{1-c}/2$ *if*

$$
t \ge \frac{1}{4\alpha}\log\log\frac{\operatorname{vol} G}{\min_x d_x} + \frac{c}{\lambda}.
$$

12.3. A comparison theorem for the log-Sobolev constant

An improvement of the convergence bounds in (12.4) depends on knowing (or estimating) the value of α, which, if anything, is harder to estimate than λ_1 (for general graphs). We can bypass this difficulty to some extent by the following (companion) comparison theorem for α. Its statement (and proof) is in fact quite close to that of Lemma 4.14.

LEMMA 12.7. *Suppose $G = (V,E)$ and $G' = (V',E')$ are connected graphs, with logarithmic Sobolev constants $\alpha = \alpha_G$ and $\alpha' = \alpha_{G'}$, respectively. Suppose $\rho : V \to V'$ is a surjective map satisfying:*

(i) *If d_x and $d'_{x'}$ denote the degrees of $v \in V$ and $x' \in V'$, respectively, then for all $x' \in V'$ we have*

$$
\sum_{x\in\rho^{-1}(x')} d_x \ge cd'_{x'} .
$$

(ii) *For each edge $e = xy \in E$ there is a path $P(e)$ between $\rho(x)$ and $\rho(y)$ in E' such that:*
    (a) *The number of edges of $P(e)$ is at most ℓ;*

(b) *For each edge $e' \in E'$, we have*

$$\sum_{e: e' \in P(e)} w_e \leq m \, w_{e'} \, .$$

where w_e denotes the weight of the edge e. Then

(12.18)
$$\alpha' \geq \frac{c}{\ell m} \alpha \, .$$

PROOF. Consider a function $g : V' \to \mathbb{R}$ which achieves the log-Sobolev constant α' in G'. Define $f : V \to \mathbb{R}$ such that $f(x) = g(\rho(x))$. Then we have

(12.19)
$$\begin{aligned}
\alpha' &= \frac{\sum\limits_{e' \in E'} g^2(e') w_{e'}}{S(g)} \\
&= \frac{\sum\limits_{e' \in E'} g^2(e') w_{e'}}{\sum\limits_{e \in E} f^2(e) w_e} \cdot \frac{\sum\limits_{e \in E} f^2(e) w_e}{S(f)} \cdot \frac{S(f)}{S(g)} \\
&= I \times II \times III \, .
\end{aligned}$$

where

$$S(g) = \sum_x f^2(x) d_x \log \frac{g(x)}{\sum\limits_z f^2(z) \pi(z)} \, .$$

Exactly as in the proof of Lemma 4.14, we obtain

$$I \geq \frac{1}{\ell m}, \; II \geq \alpha \, .$$

It remains to show $III \geq c$ (which we do using a nice idea of Holley and Stroock; cf. [**101**]). First, define

$$F(\xi, \zeta) := \xi \log \xi - \xi \log \zeta - \xi + \zeta$$

for all $\xi, \zeta > 0$. Note that $F(\xi, \zeta) \geq 0$ and for $\zeta > 0$, $F(\xi, \zeta)$ is convex in ξ. Thus, for some $c_0 > 0$,

$$\begin{aligned}
S(f) &= \sum_{x \in V} F(f^2(x), c_0) d_x \\
&= \sum_{x' \in V'} \left(\sum_{x \in \rho^{-1}(x')} d_x \right) F(g(x')^2) \\
&\geq \sum_{x' \in V'} c d'_{x'} F(g(x')^2) \text{ since } F \geq 0 \\
&\geq c \sum_{x' \in V'} F(g(x')^2 d'_{x'}) \text{ by convexity} \\
&= c S(g) \, .
\end{aligned}$$

This implies $III \geq c$ and (12.18) is proved. $\square$

If G and G' are regular with degrees k and k', respectively, then we have

$$\alpha' \geq \frac{k}{k'\ell m}\alpha .$$

12.4. Logarithmic Harnack inequalities

In previous sections we discuss log-Sobolev techniques that were adapted from the continuous case. In this section, we will consider the logarithmic Harnack inequality which was first motivated by the discrete problem on random walks. As it turns out, we can obtain the logarithmic Harnack inequalities for both Riemannian manifolds and for finite graphs.

For a smooth, compact, connected Riemannian manifold M, we let ∇ denote the gradient with the associated Laplace-Beltrami operator Δ. Suppose M has no boundary or has a boundary which is convex (as defined in (12.5)). We consider Δ acting on functions $f : M \to \mathbb{R}$ satisfying

$$(12.20) \qquad \int |f|^2 = \text{vol } M$$

satisfying the Dirichlet or Neumann boundary conditions. The log-Sobolev constant α of M is the largest value satisfying:

$$(12.21) \qquad \int_M |\nabla f(x)|^2. \geq \alpha \int_M f^2(x) \log f^2(x).$$

Suppose f is a function achieving the log-Sobolev constant α. Then it can be shown that f satisfies the following equation:

$$(12.22) \qquad \Delta f = -\alpha f \log f^2$$

Using (12.22), the following logarithmic Harnack inequality for the function f defined on M satisfying (12.20):

$$(12.23) \qquad |\nabla f|^2 \leq \max \{\alpha U^2 \log U^2, \alpha/e\}.$$

where $U = \sup |f|$ and e is the usual base of the natural logarithm, provided M has non-negative Ricci curvature.

The inequality in (12.22) is similar to the Harnack inequality except for a logarithmic factor. So, we call (12.22) the *logarithmic Harnack inequality*. It can be used to derive the following lower bound for log-Sobolev constants for a d-dimensional compact Riemannian manifold M with non-negative curvature (also see [**96**]):

$$(12.24) \qquad \alpha \geq \frac{1}{9d\, D^2(M)}$$

where $D(M)$ denotes the diameter of M.

Suppose S denote an induced subgraph of G. We consider all nontrivial functions $f : S \cup \delta S \to \mathbb{R}$ satisfying

$$\sum_x f^2(x)d_x = \text{vol } S.$$

where d_x denotes the degree of x. The log-Sobolev constant α with the Neumann boundary condition satisfies:

$$\alpha_S = \inf_f \frac{\displaystyle\sum_{\{x,y\}\in S^*}(f(x)-f(y))^2}{\displaystyle\sum_{x\in S}f^2(x)d_x \log f^2(x)}$$

where f ranges over all nontrivial functions satisfying the Neumann (or Dirichlet) boundary condition. (The case with Dirichlet boundary condition can be worked out in a similar way.)

The function f achieving the log-Sobolev constant α in (12.4) satisfies:

$$(12.25) \qquad \sum_{\substack{y \\ y\sim x}}(f(x)-f(y)) = \alpha d_x f(x) \log f^2(x)$$

where $y \sim x$ means y is adjacent to x.

For the discrete case, we can only establish the logarithmic Harnack inequality for invariant homogeneous graphs (which are defined later in Section 3).

$$(12.26) \qquad \sum_{\substack{y \\ y\sim x}}(f(x)-f(y))^2 \le \alpha d_x \max\{U^2(\log U^2 + \log U), 1\}$$

where $U = \sup_x f(x)$

For a graph G with isoperimetric dimension δ (see definition in Section 11.1) and with the assumption that G is a k-regular invariant homogeneous graph or a strongly convex subgraph, we can use (12.26) to show that

$$(12.27) \qquad \alpha \ge \frac{1}{4kD^2\delta \log \delta}$$

where D denotes the diameter of G.

Since a random walk on a graph G on n vertices is close to stationarity after order $\log\log n/\alpha$ steps, the above lower bounds for the log-Sobolev constant α immediately imply a convergence bound of order $(\log\log n)kD^2$ if the isoperimetric dimension is bounded.

The proofs for log-Harnack inequalities are somewhat complicated. We will give a proof to (12.25) and describe both the discrete and continuous results.

THEOREM 12.8. *For a graph G, suppose $f : V \to \mathbb{R}$ satisfies (12.20) and achieves the log-Sobolev constant, and $\sum_x f^2(x)d_x = \text{vol } G$. Then f satisfies, for any vertex x,*

$$Lf(x) = \sum_{\substack{y \\ y\sim x}}(f(x)-f(y)) = \alpha f(x)d_x \log f^2(x).$$

PROOF. We use Lagrange's method, taking the derivative with respect to $f(x)$ of the log-Rayleigh quotient (which is the right side of (12.4)). Then we have

$$\frac{2Lf(x)}{\displaystyle\sum_x f^2(x)\log f^2(x)} - \frac{2(f(x)d_x\log f^2(x) + 2f(x)d_x)\displaystyle\sum_{x\sim y}(f(x)-f(y))^2}{\left(\displaystyle\sum_x f^2(x)d_x\log f^2(x)\right)^2} + c_1 f(x) = 0$$

(12.28)

for some constant c_1. After substituting for α, the above expression can be simplified to:

(12.29) $\qquad Lf(x) - \alpha(f(x)\log f^2(x) + 2f(x))d_x + c_2 f(x)d_x = 0.$

After multiplying (12.29) by $f(x)$ and summing over all x in V, we have

$$\sum_{x\sim y}(f(x)-f(y))^2 - \alpha\sum_x f^2(x)d_x(\log f^2(x)+2) + c_2\sum_x f^2(x)d_x = 0.$$

This implies $c_2 = 2\alpha$. Therefore we obtain from (12.29)

$$Lf(x) = \alpha f(x)d_x\log f^2(x).$$

$\square$

We state here the following logarithmic Harnack inequality for homogeneous graphs which are invariant (as defined earlier in Chapter 9).

THEOREM 12.9. *In an invariant homogeneous graph Γ with edge generating set $\mathcal{K}$ consisting of k generators, suppose a function $f : V(\Gamma) \to \mathbb{R}$ achieves the log-Sobolev constant and satisfies (12.20). Then the following inequality holds for all $x \in V(\Gamma)$:*

$$\sum_{a\in\mathcal{K}}[f(x)-f(ax)]^2 \le 4k\sigma U^2\log U^2$$

where

$$U = \sup_y |f(y)|.$$

THEOREM 12.10. *In an abelian homogeneous graph Γ with edge generating set $\mathcal{K}$ consisting of k generators, consider a strongly convex subgraph S of Γ. Suppose a function $f : S \cup \delta S \to \mathbb{R}$ satisfies the Dirichlet or Neumann boundary condition and achieves the log-Sobolev constant. Also, assume $\displaystyle\sum_{x\in S} f^2(x)d_x = \mathrm{vol}\, S$. (This is just a scaling assumption.) Then the following inequality holds for all $x \in S$:*

$$\sum_{a\in\mathcal{K}}[f(x)-f(ax)]^2 \le 4k\sigma U^2\log U^2.$$

The proofs of Theorems 12.9 and 12.10 are similar to, but somewhat more complicated than, those in Sections 9.3, 9.4, and 9.5. The complete proofs can be found in [**62**].

There are several useful properties that a function achieving the log-Sobolev constant possesses. Here we state some without giving the proofs (see [**62**]).

THEOREM 12.11. *In a connected invariant homogeneous graph Γ with edge generating set $\mathcal{K}$ consisting of k generators, suppose a function $f : V(\Gamma) \rightarrow \mathbb{R}$ achieves the log-Sobolev constant. Then for all $x \in V(\Gamma)$, we have*

$$U = \sup_{y} |f(y)| \leq e^{k}.$$

where e is the base of the natural logarithm.

THEOREM 12.12. *In a connected invariant homogeneous invariant graph $G = (V, E)$, suppose a function $f : V \rightarrow \mathbb{R}$ satisfies the logarithmic Harnack inequality and $\sum_{x} f^2(x)d_x = \text{vol } G$. Then the log-Sobolev constant α of G satisfies*

$$\alpha \geq \min(\frac{\lambda}{4}, \frac{c}{kD^2 \log^2 U^2}),$$

where

$$U = \sup_{z} |f(z)|,$$

$$k = max_z d_z,$$

λ_1 denotes the first eigenvalue of the Laplacian of G, c denotes some absolute constant, and D denotes the diameter of G.

THEOREM 12.13. *Let S denote a strongly convex subgraph of a connected abelian homogeneous graph. Suppose a function $f : V \rightarrow \mathbb{R}$ satisfies the Dirichlet or Neumann boundary condition and achieves the log-Sobolev constant α. Also, assume $\sum_{x \in S} f^2(x)d_x = \text{vol } S$. Then the log-Sobolev constant α of G satisfies*

$$\alpha \geq \min(\frac{\lambda}{4}, \frac{c}{kD^2 \log U^2}),$$

where

$$U = \sup_{z} |f(z)|,$$

c denotes some absolute constant, k is the degree, and D denotes the diameter of S, and λ_1 denotes the first eigenvalue of the Laplacian of G.

A k-regular abelian homogenous graph or a strongly convex subgraph has the eigenvalue bound

$$\lambda_1 \geq \frac{c}{kD^2}$$

for some absolute constant c, and this lower bound is sharp up to a constant factor (the factor of k is necessary for some homogeneous graphs). As a consequence of Theorems 12.4 , 12.9, and 12.10, the log-Sobolev constant and the eigenvalue λ_1 can differ by at most a factor of $\log U$.

For graphs with isoperimetric dimension δ (defined in Section 11.1), we can derive a lower bound for α in terms of δ.

THEOREM 12.14. *Let S denote a strongly convex subgraph of a connected invariant homogeneous graph with isoperimetric dimension δ. Suppose a function $f : V \to \mathbb{R}$ satisfies the Dirichlet or Neumann boundary condition and achieves the log-Sobolev constant. Also, assume $\sum_{x \in S} f^2(x) d_x = \text{vol } S$. Then the log-Sobolev constant α of G satisfies*

$$\alpha \geq \min(\frac{\lambda}{4}, \frac{c}{kD^2 \delta \log \delta}),$$

where

$$U = \sup_z |f(z)|,$$

c denotes some absolute constant, k is the degree, and D denotes the diameter of S, and λ_1 denotes the first eigenvalue of the Laplacian of G.

Let M be a smooth connected compact Riemannian manifold and Δ be the Laplace operator associated with the Riemannian metric (as defined in Section 3.4). If the manifold M has a non-trivial boundary ∂M, we require that ∂M is convex (as defined in Chapter 10). Also, assume either the Dirichlet or Neumann boundary condition is satisfied.

THEOREM 12.15. *If f achieves the log-Sobolev constant α and $\int_M f^2 = \text{vol } M$, then f satisfies*

(12.30) $$\Delta f = -\alpha f \log f^2.$$

THEOREM 12.16. *Suppose M is a d-dimensional connected compact Riemannian manifold with non-negative Ricci curvature. If the function f achieves the log-Sobolev constant with $\int_M f^2 = \text{vol } M$, then we have*

$$\sup f \leq e^{d/2}.$$

THEOREM 12.17. *Suppose M is a connected compact Riemannian manifold with non-negative Ricci curvature that is either boundaryless or has a convex boundary. Suppose the function f achieves the log-Sobolev constant (under the Dirichlet or Neumann boundary condition) with $\int_M f^2 = \text{vol } M$. Then we have*

$$|\nabla f|^2 \leq \max\{\alpha U^2 \log U^2, \alpha/e\}$$

where

$$U = \sup f$$

and e is the base of the natural logarithm.

THEOREM 12.18. *Suppose a smooth connected compact d-dimensional manifold with non-negative Ricci curvature having no boundary or a convex boundary. Then the log-Sobolev constant α of M (with Dirichlet or Neumann condition) satisfies*

$$\alpha \geq \frac{1}{9d \, D^2(M)}$$

where $D(M)$ denotes the diameter of M.

12.5. The isoperimetric dimension and the Sobolev inequality

We recall that a graph G has *isoperimetric dimension* δ with an *isoperimetric constant* c_δ if for every subset X of $V(G)$, the total sum of edge weights between X and the complement $\bar{X}$ of X satisfies

$$\sum_{x \in X, y \notin X} w_{x,y} \geq c_\delta (\text{vol } X)^{\frac{\delta-1}{\delta}}$$

where vol $X \leq$ vol $\bar{X}$ and c_δ is a constant depending only on δ.

We will establish the following convergence bound using the isoperimetric dimension. In a weighted graph G with isoperimetric dimension δ and isoperimetric constant c_δ, we have $\Delta(t) \leq e^{-c}$ if

$$t \geq \frac{c}{\lambda \delta} + \frac{c'}{\lambda}$$

where $c' = \log(c'' c_\delta^{-1})$ for some absolute constant c''.

To prove this, we will need the following version of the Sobolev inequality (see Section 11.3 and [**60**]): For any function $g : V(G) \to \mathbb{R}$ and for $\delta > 2$,

$$(12.31) \qquad \sum_{u \sim v} |g(u) - g(v)|^2 w(x,y) \geq c \inf_m (\sum_v |g(v) - m|^{2\gamma} d_v)^{\frac{1}{\gamma}}$$

where $\gamma = \frac{\delta}{\delta-2}$ and $c = c_\delta (\delta - 1)^2 / 4\delta^2$.

We will prove a discrete version of the classical regularity theory of De-Georgi-Nash-Moser [**254**]:

THEOREM 12.19. *For a weighted graph G with isoperimetric dimension δ, suppose f is a harmonic eigenfunction with eigenvalue λ. Then we have*

$$\sup_x | f^2(x) | \leq c^{\delta/2} \lambda^{\delta/2} \sum_x f^2(x) d_x$$

for some absolute constant c.

PROOF. By (12.13), we have

$$\sum_x \lambda |f^p(x)| d_x = \sum_x |f^{p-1}(x)| \, L \, f(x)$$

$$= \sum_{x \sim y} (f(x) - f(y)) \cdot (f^{p-1}(x) - f^{p-1}(y)) w(x,y)$$

$$(12.32) \qquad \geq \sum_{x \sim y} \frac{4(p-1)}{p^2} \left(f^{p/2}(x) - f^{p/2}(y) \right)^2 w(x,y).$$

Now we use the Sobolev inequality (12.31) with $g = f^{p/2}$ and also (12.32). We have

$$\left(\sum_x f^{\gamma p}(x) d_x \right)^{1/\gamma} \leq c^{-1} \sum_{x,y} \left(f^{p/2}(x) - f^{p/2}(y) \right)^2 w(x,y)$$

$$\leq c^{-1} \frac{p^2}{4(p-1)} \lambda \sum_x f^p(x) d_x.$$

Or, for $q \geq 0$,

$$(12.33) \qquad \left(\sum_x f^{\gamma q}(x) d_x \right)^{1/\gamma} \leq c^{-1} \frac{q^2}{q-1} \lambda \sum_x f^q(x) d_x.$$

We apply (12.33) recursively. For $q = 2$, we have

$$\left(\sum_x f^{2\gamma}(x) d_x \right)^{1/\gamma} \leq c_1 4\lambda \sum_x f^2(x) d_x$$

where $c_1 = c^{-1}$. Setting $q = 2\gamma$ we have

$$\left(\sum_x f^{2\gamma^2}(x) d_x \right)^{1/\gamma^2} \leq \left(c_1 \frac{4\gamma^2}{2\gamma - 1} \lambda \sum_x f^{2\gamma}(x) d_x \right)^{1/\alpha}$$

$$\leq (c_1 \frac{4\gamma^2}{2\gamma - 1} \lambda)^{1/\gamma} \left(\sum_x f^{2\gamma}(x) d_x \right)^{1/\gamma}$$

$$\leq (c_1 \lambda)^{1+1/\gamma} 4 \left(\frac{4\gamma^2}{2\gamma - 1} \right)^{1/\gamma} \sum_x f^2(x) d_x.$$

Extending to $q = 2\gamma^i$, we get

$$\left(\sum_x f^{2\gamma^i}(x) d_x \right)^{1/\gamma^i}$$

$$\leq (c_1 \lambda)^{1+1/\alpha+\cdots+1/\gamma^i} 4 \left(\frac{4\gamma^2}{2\gamma - 1} \right)^{1/\gamma} \left(\frac{4\gamma^4}{2\gamma^2 - 1} \right)^{1/\gamma^2} \cdots \sum_x f^2(x) d_x$$

We now note that

$$1 + \frac{1}{\gamma} + \cdots + \frac{1}{\gamma^i} \leq \frac{1}{1 - \frac{1}{\gamma}} = \frac{\delta}{2}.$$

Moreover,

$$4 \left(\frac{4\gamma^2}{2\gamma - 1} \right)^{1/\gamma} \left(\frac{4\gamma^4}{2\gamma^2 - 1} \right)^{1/\gamma^2} \cdots \leq 4^{\delta/2} \gamma^{\frac{1}{\gamma} + \frac{2}{\gamma^2} + \cdots + \frac{i}{\gamma^i}}$$

$$\leq 4^{\delta/2} \gamma^{\delta/(2\gamma - 2)}$$

$$\leq (4e^2)^{\delta/2}.$$

Therefore we have

$$\sup_x | f^2(x) | \leq c_2^{\delta/2} \lambda^{\delta/2} \sum_x f^2(x) d_x$$

where $c_2 = 4e^2 c_1$. The proof of Theorem 12.19 is complete. $\square$

We note that in the proof of Theorem 12.19, it suffices to start with a function f satisfying (12.32). Here we state a slightly more general version of Theorem 12.19.

COROLLARY 12.20. *For a weighted graph G with isoperimetric dimension δ, suppose f is a function satisfying*

$$\sum_x \lambda f^p(x) d_x \geq \sum_x f^{p-1}(x) \ L \ f(x).$$

Then we have

$$\sup_x \mid f^2(x) \mid \leq c^{\delta/2} \ \lambda^{\delta/2} \sum_x f^2(x) d_x$$

for some absolute constant c.

As a consequence of the above theorem, we have

THEOREM 12.21. *In a weighted graph G with isoperimetric dimension δ and isoperimetric constant c_δ, the random walk approaches stationarity in $O(\dfrac{\log \operatorname{vol} G}{\lambda \delta} + \dfrac{c'}{\lambda})$ steps where $c' = \log(c'' c_\delta^{-1})$ for some absolute constant c'' and $\lambda = \lambda_1$ if $1 - \lambda_1 \geq \lambda_{n-1} - 1$ and $\lambda = 2\lambda_1/(\lambda_1 + \lambda_{n-1})$ otherwise.*

PROOF. We replace f by $g_x = \psi_x \pi^{-1/2}(H_t - I_0)\pi^{-1/2}$ in Theorem 12.19. It can be checked that

$$
\begin{aligned}
\sum_y g_x^2(y) d_y &= \operatorname{vol} G \ \psi_x \pi^{-1/2}(H_{2t} - I_0)\pi^{-1/2}\psi_x \\
&\leq e^{-2t\lambda}\|\pi^{-1/2}\psi_x\|^2 \operatorname{vol} G \\
&\leq e^{-2t\lambda}\frac{\operatorname{vol} G}{d_x}.
\end{aligned}
$$

Since g_x is orthogonal to the eigenfunction ϕ_0 of $\mathcal{L}$, we have

$$e^{-t\lambda} \sum_y g_x^p(y) d_y \geq \sum_y g_x^{p-1}(y) \ L \ g_x(y).$$

By Corollary 12.20, we have

$$\sup_x \mid g_x^2(x) \mid \leq c^{\delta/2} e^{-\lambda t \delta/2} \sum_y g_x^2(y) d_y.$$

Finally, we consider

$$
\begin{aligned}
\Delta(t) &= \max_{x,y} \frac{|P^t(y,x) - \pi(x)|}{\pi(x)} \\
&= \max_x \max_y |g_x(y)| \\
&\leq \max_x c^{\delta/2} e^{-t\lambda\delta/2} \sum_y g_x^2(y) d_y \\
&\leq c^{\delta/2} e^{-t\lambda\delta/2} \frac{\operatorname{vol} G}{\min_x d_x}
\end{aligned}
$$

where $c = c'' c_\delta^{-1}$. Theorem 12.21 is proved.   □

Many invariants of the hypercube Q_n can be effectively evaluated using the fact that Q_n is the cartesian product of an edge. The convergence bound for the random walk problem on Q_n is of order $n \log n$ (as described at the end of Chapter 1).

When dealing with subgraphs of Q_n, the random walk problems become more complicated. We say Q'_n is a punctured cube if it is a subgraph of Q_n obtained by removing one vertex from Q_n. We will use the isoperimetric dimension approach as follows:

The isoperimetric problems on hypercubes have long been studied. For a fixed number m, the edge boundary of a subset $X \subseteq V(Q_n)$ with vol $X = m$, is minimized when X is close to a *Hamming ball* (i.e., $N_r(v)$ for some v and r). In other words, the isoperimetric dimension for Q_n is $\delta = n/\log n$ and the isoperimetric constant $c_\delta = 1$. Using Theorem 12.21, we have

$$\Delta(t) \leq \frac{\log \text{ vol } Q_n}{\delta \lambda}$$
$$\leq n \log n.$$

For a punctured cube Q'_n, the isoperimetric dimension stays virtually unchanged. So it follows from Theorem 12.21, that its (lazy) random walk converges in time $O(n \log n)$.

Notes

In Chapter 9 and 12, we have given proofs for Harnack inequalities and logarithmic Harnack inequalities for invariant homogeneous graphs and their convex subgraphs. These results can be generalized by considering *Ricci flat* graphs, which we will define below:

In a graph G with vertex set $V = V(G)$ and edge set $E = E(G)$, the neighborhood $N^*(v)$ of vertex v consists of v and vertices adjacent to v. We say G has a *local k-frame* at v if there are mappings $\eta_1, \cdots, \eta_k \colon N^*(v) \to V$ satisfying
(1) G is k-regular;
(2) u is adjacent to $\eta_i u$ for every $u \in V$ and $1 \leq i \leq k$;
(3) $\eta_i u \neq \eta_j u$ if $i \neq j$.

A graph G is said to be *Ricci flat* if G has a local k-frame and

$$\bigcup_j \eta_i \eta_j v = \bigcup_j \eta_j \eta_i v$$

for any i and v.

For example, a homogeneous graph associated with an abelian group is Ricci flat [58]. An invariant homogeneous graph is Ricci flat. For more discussions on Ricci flat graphs, see [62].

We remark that Theorems 9.4, 9.5, 9.7, 9.8, 9.9, 9.10, 9.11, 12.9, 12.10, 12.11, 12.12, 12.13, 12.14 can all be generalized by replacing invariant homogeneous graphs or abelian homogeneous graphs by Ricci flat graphs using very similar proofs.

Bibliography

[1] H.L. Abbott, Lower bounds for some Ramsey numbers, *Discrete Math.* **2** (1972), 289-293.

[2] A. Agresti, *Categorical Data Analysis*, John Wiley and Sons, New York, 1990.

[3] D. Aldous, On the Markov-chain simulation method for uniform combinatorial simulation and simulated annealing, *Prob. Eng. Info. Sci.* **1** (1987), 33-46.

[4] D. Aldous, Some inequalities for reversible Markov chains, *J. London Math. Soc.* **25** (1982), 564-576.

[5] N. Alon, Eigenvalues and expanders, *Combinatorica* **6** (1986), 86-96.

[6] N. Alon and F.R.K. Chung, Explicit constructions of linear-sized tolerant networks, *Discrete Math.* **72** (1988), 15-20.

[7] N. Alon, F.R.K. Chung and R.L. Graham, Routing permutations on graphs via matchings, *SIAM J. Disc. Math.* **7** (1994), 513-530.

[8] N. Alon, Z. Galil and V.D. Milman, Better expanders and superconcentrators, *J. Algorithms* **8** (1987), 337-347.

[9] N. Alon and V. D. Milman, λ_1 isoperimetric inequalities for graphs and superconcentrators, *J. Comb. Theory* B 38 (1985), 73-88.

[10] N. Alon, Z. Galil and O. Margalit, On the exponent of the all pairs shortest paths problem, *32nd Symposium on Foundations of Computer Science*, IEEE Computer Society Press, (1991), 569-575.

[11] N. Alon and N. Kahale, Approximating the independence number via the θ function, *Math. Programming*, to appear.

[12] N. Alon and J.H. Spencer, *The Probabilistic Method*, John Wiley and Sons, New York 1991.

[13] S. Arora and S. Safra, Probabilistic checking of proofs; a new characterization of NP. *33nd Symposium on Foundations of Computer Science*, IEEE Computer Society Press, (1992), 2-13.

[14] L. Babai and M. Szegedy, Local expansion of symmetrical graphs, *Combinatorics, Probability and Computing* 1 (1991), 1-12.

[15] L. Babai, Automorphism groups, isomorphism, reconstruction, *Handbook of Combinatorics* (eds. R. L. Graham, Grötschel and L. Lovász), North-Holland, Amsterdam, (1996), 1447-1540.

[16] D. Bakry, L'hypercontractivité et son utilisation en théorie des semigroups, In *Ecole d' été de Saint Fleur 1992*, Springer Lecture Notes 1581.

[17] L.A. Bassalygo, Asymptotically optimal switching circuits, *Problems Inform. Transmission* **17** (1981), 206-211.

[18] J. Beck, On size Ramsey number of paths, trees and circuits I., *J. Graph Theory* **7** (1983), 115-129.

[19] W. Beckner, Inequalities in Fourier analysis, *Annals of Mathematics* **102** (1975), 159-182.

[20] V.E. Beněs, Mathematical Theory of Connecting Networks, Academic Press, New York 1965.

[21] J.C. Bermond and B. Bollobás, The diameter of graphs – a survey, *Congressus Numerantium* **32** (1981), 3-27.

[22] J.C. Bermond, C. Delorme and G. Farhi, Large graphs with given degree and diameter, III, *Proc. Coll. Cambridge* (1981), *Ann. Discr. Math.* **13**, North Holland, Amsterdam (1982), 23-32.

[23] A. J. Berstein, Maximally connected arrays on the n-cube, *SIAM J. Appl. Math.* **15** (1967), 1485-1489.

[24] F. Bien, Constructions of telephone networks by group representations, *Notices Amer. Math. Soc.* **36** (1989), 5-22.

[25] N.L. Biggs, *Algebraic Graph Theory*, (2nd ed.), Cambridge University Press, Cambridge, 1993.

[26] N.L. Biggs and M.H. Hoare, The sextet construction for cubic graphs, *Combinatorica* **3** (1983), 153-165.

[27] N.L. Biggs, E.K. Lloyd and R.J. Wilson, *Graph Theory 1736-1936*, Clarendon Press, Oxford, 1976.

[28] Y. Bishop, S. Fienberg, P. Holland, *Discrete Multivariate Analysis*, MIT Press, Cambridge, 1975.

[29] M. Blum, R.M. Karp, O.Vornberger, C. H. Papadimitriou, and M. Yannakakis, The complexity of testing whether a graph is a superconcentrator, *Inf. Proc. Letters* **13** (1981), 164-167.

[30] B. Bollobás, *Random Graphs*, Academic Press, New York (1987).

[31] B. Bollobás, *Extremal Graph Theory*, Academic Press, London (1978).

[32] B. Bollobás and F.R.K. Chung, The diameter of a cycle plus a random matching,*SIAM J. on Discrete Mathematics* **1** (1988), 328-333.

[33] B. Bollobás and I. Leader, Edge-isoperimetric inequalities in the grid, *Combinatorica* **11**(1991), 299-314.

[34] B. Bollobás and I. Leader, An isoperimetric inequality on the discrete torus, *SIAM J. Disc. Math.* **3** (1990), 32-37.

[35] B. Bollobás, and A. Thomason, Graphs which contain all small graphs, *European J. of Combinatorics* **2** (1981), 13-15.

[36] B. Bollobás and W.F. de la Vega, The diameter of random graphs, *Combinatorica* **2** (1982), 125-134.

[37] J.A. Bondy and M. Simonovits, Cycles of even length in graphs, *J. Combin. Theory Ser. B* **16** (1974), 97-105.

[38] R.B. Boppana, Eigenvalues and graph bisection: An average-case analysis, *28nd Symposium on Foundations of Computer Science*, IEEE Computer Society Press, (1987), 280-285.

[39] R. Bott and J.P. Mayberry, Matrices and trees, In *Economic Activity Analysis*, (O. Morgenstern, ed.), John Wiley and Sons, New York (1954), 391-340.

[40] A. Broder, A. Frieze and E. Upfal, Existence and construction of edge disjoint paths on expander graphs, *Proc. Sym. Theo. on Computing*, ACM (1992), 140-149.

[41] R. Brooks, The spectral geometry of k-regular graphs, Journal d'Analyse Mathématique, **57** (1991), 120-151.

[42] N.G. de Bruijn, A combinatorial problem, *Nederl. Akad. Wetensch. Proc.* **49** (1946), 758-764.

[43] D.A. Burgess, On character sums and primitive roots, *Proc. London Math. Soc.* **12** (1962) 179-192.

[44] P. Buser, Cayley graphs and planar isospectral domains, in *Geometry and Analysis on Manifolds* (T. Sunada, ed.), Springer Lecture Notes 1339 (1988), 64-77.

[45] P. Buser, Cubic graphs and the first eigenvalue of a Riemann surface, *Math. Z.* **162** (1978), 87-99.

[46] L. Caccetta, On extremal graphs with given diameter and connectivity, *Ann. New York Acad. Sci.* **328** (1979), 76-94.

[47] A. Cayley, A theorem on trees, *Quart. J. Math.* 23 (1889), 376-378.

[48] J. Cheeger, A lower bound for the smallest eigenvalue of the Laplacian, Problems in Analysis (R. C. Gunning, ed.), Princeton Univ. Press (1970), 195-199.

[49] Siu Yuen Cheng, Peter Li and Shing-Tung Yau, On the upper estimate of the heat kernel of a complete Riemannian manifold, *American Journal of Mathematics* **103** (1981), 1021-1063.

[50] R. Christensen, *Log-Linear Models*, Springer-Verlag, New York, 1990.

[51] F.R.K. Chung, Diameters and eigenvalues, *J. of Amer. Math. Soc.* **2** (1989), 187-196.

[52] F.R.K. Chung, Eigenvalues of graphs and Cheeger inequalities, in *Combinatorics, Paul Erdős is Eighty*, Volume 2, edited by D. Miklós, V. T. Sós, and T. Szőnyi, János Bolyai Mathematical Society, Budapest (1996), 157-172.

[53] F.R.K. Chung, V. Faber and T. A. Manteuffel, On the diameter of a graph from eigenvalues associated with its Laplacian, *SIAM. J. Discrete Math.* **7**(1994), 443-457.

[54] F.R.K. Chung, A. Grigor'yan, and S.-T. Yau, Upper bounds for eigenvalues of the discrete and continuous Laplace operators, *Advances in Mathematics* **117** (1996), 165-178.

[55] F.R.K. Chung, A. Grigor'yan, and S.-T. Yau, Eigenvalues and diameters for manifolds and graphs, *Tsing Hua Lectures on Geometry and Analysis*, to appear.

[56] F.R.K. Chung and S.-T. Yau, A Harnack inequality for Dirichlet eigenvalues, preprint.

[57] F.R.K. Chung and Prasad Tetali, Isoperimetric inequalities for cartesian products of graphs, *Combinatorics, Probability and Computing*, to appear.

[58] F.R.K. Chung and S.-T. Yau, A Harnack inequality for homogeneous graphs and subgraphs, *Communications in Analysis and Geometry*, **2** (1994), 628-639.

[59] F.R.K. Chung and S.-T. Yau, Eigenvalues of graphs and Sobolev inequalities, *Combinatorics, Probability and Computing*, **4** (1995), 11-26.

[60] F.R.K. Chung and S.-T. Yau, Eigenvalue inequalities of graphs and convex subgraphs, *Communications in Analysis and Geometry*, to appear.

[61] F.R.K. Chung and S.-T. Yau, Eigenvalues, flows and separators of graphs, preprint.

[62] F.R.K. Chung and S.-T. Yau, Logarithmic Harnack inequalities, *Mathematics Research Letters*, **3**, (1996), 793-812.

[63] F.R.K. Chung, On concentrators, superconcentrators, generalizers and nonblocking networks, *Bell Systems Tech. J.* **58** (1978), 1765-1777.

[64] F.R.K. Chung, A note on constructive methods for Ramsey numbers, *J. Graph Th.* **5** (1981), 109-113.

[65] F.R.K. Chung, Diameters of communications networks, *Mathematics of Information Processing*, AMS Short Course Lecture Notes (1984), 1-18.

[66] F.R.K. Chung, Diameters of graphs: Old problems and new results, *Congressus Numerantium* **60** (1987), 295-317.

[67] F.R.K. Chung, Quasi-random classes of hypergraphs, *Random Structures and Algorithms* **1** (1990), 363-382.

[68] F.R.K. Chung and M.R. Garey, Diameter bounds for altered graphs, *J. of Graph Theory* **8** (1984), 511-534.

[69] F.R.K. Chung, Constructing random-like graphs, in *Probabilistic Combinatorics and Its Applications*, (B. Bollobas ed.), Amer. Math. Soc., Providence, (1991) 21-56

[70] F.R.K. Chung and R.L. Graham, Quasi-random hypergraphs, *Random Structures and Algorithms* **1** (1990), 105-124.

[71] F.R.K. Chung and R.L. Graham, Quasi-random tournaments, *J. of Graph Theory* **15** (1991), 173-198.

[72] F.R.K. Chung and R.L. Graham, Maximum cuts and quasi-random graphs, *Random Graphs* (Alan Frieze and Tomasz Luczak, eds.), John Wiley and Sons, New York (1992), 23-34.

[73] F.R.K. Chung and R.L. Graham, On graphs not containing prescribed induced subgraphs, in *A Tribute to Paul Erdös*, (A. Baker et al. eds.) Cambridge University Press (1990), 111-120.

[74] F.R.K. Chung and R.L. Graham, Quasi-random set systems, *J. Amer. Math. Soc.* **4** (1991), 151-196.

[75] F.R.K. Chung and R.L. Graham, Quasi-random subsets of Z_n, *J. Combin. Th. (A)* **61** (1992), 64-86.

[76] F.R.K. Chung, R.L. Graham and R.M. Wilson, Quasi-random graphs, *Combinatorica* **9** (1989), 345-362.

[77] F.R.K. Chung, The regularity lemma for hypergraphs and quasi-randomness, *Random Structures and Algorithms* **2** (1991), 241-252.

[78] F.R.K. Chung and R.L. Graham, Cohomological aspects of hypergraphs, *Trans. Amer. Math. Soc.* **334** (1992), 365-388

[79] F.R.K. Chung, R.L. Graham and S.-T. Yau, On sampling with Markov chains, *Random Structures and Algorithms* **9** (1996) 55-78.

[80] F.R.K. Chung and R.L. Graham, Random walks on generating sets of groups, *Electronic J. Combinatorics*, **4**, no. 2, (1997), #R7.

[81] F.R.K. Chung and R.L. Graham, Stratified random walks on an n-cube, *Random Structures and Algorithms*, to appear.

[82] F.R.K. Chung and C.M. Grinstead, A survey of bounds for classical Ramsey numbers, *J. Graph Theory* **7** (1983), 25-38.

[83] F.R.K. Chung and K. Oden, Weighted graph Laplacians and isoperimetric inequalities, preprint.

[84] F.R.K. Chung and S. Sternberg, Laplacian and vibrational spectra for homogenous graphs, *J. Graph Theory* **16** (1992), 605-627.

[85] F.R.K. Chung and S. Sternberg, Mathematics and the Buckyball, *American Scientist*, **81** (1993), 56-71.

[86] F.R.K. Chung, B. Kostant and S. Sternberg, Groups and the Buckyball, in *Lie Theory and Geometry: In honor of Bertram Kostant* (Eds. J.-L. Brylinski, R. Brylinski, V. Guillemin and V. Kac) PM 123, Birkhäuser, Boston, 1994.

[87] F.R.K. Chung, D. Rockmore, S. Sternberg, On the symmetry of the Buckyball, preprint.

[88] F.R.K. Chung and P. Tetali, Communication complexity and quasi-randomness, *SIAM J. Discrete Math.* **6** (1993), 110-123.

[89] F.R.K. Chung and R.P. Langlands, A combinatorial Laplacian with vertex weights, *J. Comb. Theory* (A), **75** (1996), 316-327.

[90] D. M. Cvetković, M. Doob and H. Sachs, *Spectra of Graphs, Theory and Application*, Academic Press, 1980.

[91] D. M. Cvetković, M. Doob, I. Gutman, and A. Torgašev, *Recent results in the Theory of Graph Spectra*, North Holland, Amsterdam 1988.

[92] E.B. Davies, Heat kernel bounds, conservation of probability and the Feller property, *J. d'Analyse Math.* **58**, (1992), 99-119.

[93] P.J. Davis, *Circulant Matrices*, John Wiley and Sons, New York (1979).

[94] P. Deligne, La conjecture de Weil I, *Inst. Hautes Etudes Sci. Publ. Math* **43** (1974) 273-307.

[95] E. D'Hoker and D.H. Phong, On determinants of Laplacians on Riemann surfaces, *Comm. Math. Phys.* **104** (1986), 537-545.

[96] J.-D. Deuschel and D.W. Stroock, Hypercontractivity and spectral gap of symmetric diffusions with applications to the stochastic Ising models, *J. Funct. Anal.* **92** (1990), 30-48.

[97] P. Diaconis, *Group Representations in Probability and Statistics*, Institute of Math. Statistics, Hayward, California, 1988.

[98] P. Diaconis and D.W. Stroock, Geometric bounds for eigenvalues of Markov chains, *Annals Applied Prob.* **1** (1991), 36-61.

[99] P. Diaconis and L. Saloff-Coste, Comparison theorems for reversible Markov chains, *Annals of Applied Prob.* **3** (1993), 696-730.

[100] P. Diaconis and L. Saloff-Coste, Comparison techniques for random walks on finite groups *Annals of Applied Prob.* **4** (1993), 2131-2156.

[101] P. Diaconis and L. Saloff-Coste, Logarithmic Sobolev inequalities for finite Markov chains, *Ann. Appl. Prob.* **6** (1996), 695-750.

[102] P. Diaconis and L. Saloff-Coste, Walks on generating sets of groups, *Prob. Theory Related Fields* **105** (1996), 393-421.

[103] P. Diaconis and B. Sturmfels, Algebraic algorithms for sampling from conditional distributions, preprint.

[104] P. Diaconis, R.L. Graham and J. Morrison, Asymptotic analysis of a random walk on a hypercube with many dimensions, *Random Structures and Algorithms* **1** (1990), 51-72.

[105] P. Diaconis and L. Saloff-Coste, An application of Harnack inequalities to random walk on nilpotent quotients, *J. Fourier Anal. Appl.* (1995), 189-207.

[106] M.J. Dinneen, M.R. Fellows and V. Faber, Algebraic constructions of efficient broadcast networks, *Lecture Notes in Comput. Sci.*, *539*, Springer, Berlin, 1991, 152-158.

[107] J. Dodziuk and L. Karp, Spectral and function theory for combinatorial Laplacians, in Geometry of Random Motion, *Contemp. Math* **73**, AMS Publication (1988), 25-40.

[108] M. Dyer, A. Frieze and R. Kannan, A random polynomial time algorithm for approximating the volume of convex bodies, *JACM* **38** (1991), 1-17.

[109] M. Eichler, Quaternary quadratic forms and the Riemann hypothesis for congruence zeta functions, *Arch. Math.* **5** (1954), 355-366.

[110] B. Elspas, Topological constraints on interconnection limited logic, *Switching Circuit Theory and Logical Design* **5** (1964), 133-147.

[111] P. Erdős, Some remarks on the theory of graphs, *Bull. Amer. Math. Soc.* **53** (1947), 292-294.

[112] P. Erdős, Some remarks on chromatic graphs, *Colloquium Mathematicum* **16** (1967), 253-256.

[113] P. Erdős, S. Fajtlowicz and A.J. Hoffman, Maximum degree in graphs of diameter 2, *Networks* **10** (1980), 87-90.

[114] P. Erdős and A. Hajnal, On spanned subgraphs of graphs, *Betrage zur Graphentheorie und deren Anwendungen*, Kolloq. Oberhof (DDR) (1977), 80-96.

[115] P. Erdős and A. Rényi, On a problem in the theory of graphs, *Publ. Math. Inst. Hungar. Acad. Sci.* **7** (1962), 623-641.

[116] P. Erdős, A. Rényi and V.T. Sós, On a problem of graph theory, *Studia Sci. Math. Hungar.* **1** (1966), 215-235.

[117] P. Erdős and H. Sachs, Reguläre Graphen gegenebener Teillenweite mit Minimaler Knotenzahl, Wiss. Z. Univ. Halle – Wittenberg, *Math. Nat. R.* **12** (1963), 251-258.

[118] P. Erdős and V.T. Sós, On Ramsey-Turán type theorems for hypergraphs, *Combinatorica* **2** (1982), 289-295.

[119] P. Erdős and J. Spencer, Imbalances in k-colorations, *Networks* **1** (1972), 379-386.

[120] P. Erdős and J. Spencer, *Probabilistic Methods in Combinatorics*, Academic Press, New York (1974).

[121] R.J. Faudree and M. Simonovits, On a class of degenerate extremal graph problems, *Combinatorica* **3** (1983), 97-107.

[122] U. Feige, Randomized graph products, chromatic numbers, and the Lovász θ-function, *Proc. Sym. Theo. on Computing*, ACM (1995), 635-640.

[123] U. Feige, S. Goldwasser, L. Lovász, S. Safra, and M. Szegedy, Approximating clique is almost NP-complete, *32nd Symposium on Foundations of Computer Science*, IEEE Computer Society Press, (1991), 2-12.

[124] P. Frankl and V. Rödl, Forbidden intersections, *Trans. AMS* **300** (1987), 259-286.

[125] J.A. Fill, Eigenvalue bounds on convergence to stationarity for nonreversible Markov chains, with an application to the exclusion process, *Ann. Appl. Prob.* **1** (1991) 62-87.

[126] M. Friedler, Algebraic connectivity of graphs, *Czech. Math. J.* **23** (98) (1973), 298-305.

[127] L.R. Ford and D.R. Fulkerson, *Flows in Networks*, Princeton Univ. Press (1962).

[128] P. Frankl, A constructive lower bound for some Ramsey numbers, *Ars Combinatoria* **3** (1977), 297-302.

[129] P. Frankl, V. Rödl and R.M. Wilson, The number of submatrices of given type in a Hadamard matrix and related results, *J. Combinatorial Th. (B)* **44** (1988), 317-328.

[130] P. Frankl and R.M. Wilson, Intersection theorems with geometric consequences, *Combinatorica* **1** (1981), 357-368.

[131] M. L. Fredman, New bounds on the complexity of the shortest path problem, *SIAM J. Computing* **5** (1976), 83-89.

[132] Joel Friedman, On the second eigenvalue and random walks in random d-regular graphs, *Combinatorica* **11** (1991), 331-362.

[133] J. Friedman and N. Pippenger, Expanding graphs contain all small trees, *Combinatorica* **7** (1987), 71-76.

[134] G.Frobenius, Über die Charaktere der alternierenden Gruppe, *Sitzungsberichte Preuss. Akad. Wiss.* (1901), 303-315.

[135] G. Frobenius, Über Matrizen aus nicht negative Elementen, *Sitzber. Akad. Wiss. Berlin* (1912), 456-477.

[136] Z. Füredi and J. Komlós, The eigenvalues of random symmetric matrices, *Combinatorica* **1** (1981), 233-241.

[137] O. Gabber and Z. Galil, Explicit construction of linear sized superconcentrators, *J. Comput. System Sci.* **22** (1981), 407-420.

[138] A. Galtman, Laplacians, the Lovász number, and Delsarte's linear programming bound, preprint.

[139] F.R. Gantmacher, *The Theory of Matrices*, Vol. 1, Chelsea Pub. Co., New York (1977).

[140] M.R. Garey and D.S. Johnson, *Computers and Intractability, A Guide to the Theory of NP-Completeness*, W. H. Freeman and Co., San Francisco, 1979.

[141] M. Goemans and D. Williamson, Improved approximation algorithms for maximum cut and satisfiability problems using semidefinite programming, J. ACM **42** (1995), 1115-1145.

[142] P. Ginsbarg, *Applied Conformal Field Theory*, les Houches (1988).

[143] W. Goddard, private communication.

[144] R.L. Graham and V. Rödl, *Numbers in Ramsey theory, Surveys in Combinatorics* (1987), (C. Whitehead, ed.) London Math. Soc. Lecture Notes Series 123, 111-153.

[145] S.W. Graham and C. Ringrose, Lower bounds for least quadratic non-residues, *Analytic Number Theory*, (edited by B. Berndt, H. Halberstam, H. Diamond and A. Hildebrand), Birkhäuser, Boston 1990.

[146] R.L. Graham and J.H. Spencer, A constructive solution to a tournament problem, *Canad. Math. Bull.* **14** (1971), 45-48.

[147] A. Grigor'yan, Integral maximum principle and its applications, *Proc. of Royal Society, Edinburgh Sect. A*, **124** (1994), 353-362.

[148] M. Gromov, Groups of polynomial growth and expanding maps, *Publ. IHES*, 53 (1981), 53-78.

[149] L. Gross, Logarithmic Sobolev inequalities, *Amer. J. Math.* **97** (1976), 1061-1083.

[150] L. Gross, *Logarithmic Sobolev inequalities and contractivity properties of semi-groups*, Springer Lecture Notes 1563. (1993).

[151] M. Grotschel, L. Lovász and A. Schrijver, The ellipsoid method and its consequences in combinatorial optimization, *Combinatorica* **1** (1981), 169-197.

[152] M. Grotschel, L. Lovász and A. Schrijver, *Geometric Algorithms and Combinatorial Optimization*, Springer-Verlag, Berlin, 1988.

[153] Q.-P. Gu and H. Tamaki, Routing a permutation in the hypercube by two sets of edge-disjoint paths, *Proc. of 10th International Parallel Processing Symposium*, IEEE Computer Society Press, (1996).

[154] W. Haemers, Eigenvalue methods in *Packing and Covering in Combinatorics*, (A. Schrijver ed.), Mathematisch Centrum, Amsterdam (1982), 15–38.

[155] M.M. Halldorsson, A still better performance quarantee for approximating graph coloring, *Information Processing Letters* **45** (1993), 19-23.

[156] L. H. Harper, Optimal numberings and isoperimetric problems on graphs, *J. of Comb. Theory* **1** (1966), 385-393.

[157] S. Hart, A note on the edges of the n-cube, *Discrete Math.* **14** (1976), 157-163.

[158] J. Hastad, Clique is hard to approximate within $n^{1-\epsilon}$, *37nd Symposium on Foundations of Computer Science*, IEEE Computer Society Press, (1996), 627-636.

[159] J. Haviland and A. Thomason, Pseudo-random hypergraphs, *Discrete Math.* **75** (1989), 255-278.

[160] W. Hebisch and L. Saloff-Coste, Gaussian estimates for Markov chains and random walks on groups, preprint.

[161] A.J. Hoffman and R.R. Singleton, On Moore graphs with diameter 2 and 3, *IBM J. of Res. Development* **4** (1960), 497-504.

[162] A. J. Hoffman, Eigenvalues of graphs, in *Studies in Graph Theory* II (D. R. Fulkerson, ed.), M.A.A. Studies in Math, Washington D.C.,(1975), 225–245.

[163] R. A. Horn and C.R. Johnson, *Matrix Analysis*, Cambridge University Press, New York, 1985.

[164] J. Igusa, Fibre systems of Jacobian varieties III, *American J. of Math.* **81** (1959), 453-476.

[165] W. Imrich, Explicit construction of regular graphs without small cycles, *Combinatorica* **4** (1984), 53-59.

[166] K. Ireland and M. Rosen, *A Classical Introduction to Modern Number Theory*, Springer-Verlag, New York 1982.

[167] M. Jerrum and A. J. Sinclair, Approximating the permanent, *SIAM J. Computing* **18** (1989), 1149-1178.

[168] F. Juhász, On the spectrum of a random graph, *Colloq. Math. Soc. János Bolyai* **25**, *Algebraic Methods in Graphs Theory*, Szeged (1978), 313-316.

[169] N. Kahale, Isoperimetric inequalities and eigenvalues, *SIAM J. Discrete Math.*, **10** (1997), 30-40.

[170] B.S. Kashin and S.V. Konyagin, On systems of vectors in a Hilbert space, *Trudy Mat. Inst. imeni V. A. Steklova* **157** (1981) 64-67. English translation in *Proceedings of the Steklow Institute of Math.* AMS (1983), 67-70.

[171] G.O.H. Katona, A theorem on finite sets, *Theory of Graphs, Proc. Colloq. Tihany*, pp. 187-207, Academic Press, New York, 1966.

[172] N.M. Katz, An estimate for character sums, *J. Amer. Math. Soc.* **2** (1989), 197-200.

[173] F. Kirchhoff, Über die Auflösung der Gleichungen, auf welche man bei der Untersuchung der linearen Verteilung galvanischer Ströme geführt wird, *Ann. Phys. chem.* **72** (1847), 497-508

[174] D. Karger, R. Motwani, and M. Sudan, Approximate graph coloring by semi-definite programming, *35th Symposium on Foundations of Computer science*, IEEE Computer Society Press, 1994, 2-13.

[175] M. Klawe, Non-existence of one-dimensional expanding graphs, *22nd Symposium on Foundations of Computer Science*, IEEE Computer Society Press, (1981), 109-113.

[176] D.E. Knuth, *The Art of Computer Programming*, vol. 3, p. 241, Addison-Wesley, Reading, MA, 1973.

[177] D.E. Knuth, The sandwich theorem, *Electronic J. of Combinatorics* **1** (1994), A1.

[178] Robert P. Langlands and Yvan Saint-Aubin, Algebro-geometric aspects of the Bethe equations, preprint.

[179] J. B. Kruskal, The number of simplices in a complex, *Mathematical Optimization Techniques*, (ed. R. Bellman), pp. 251-78, University of California Press, Berkeley, 1963.

[180] F. Lazebnik, V. A. Ustimenko and A. J. Woldar, A new series of dense graphs of high girth, *Bull. Amer. Math. Soc.* **32** (1995), 73-79.

[181] F.T. Leighton, *Introduction to Parallel Algorithms and Architectures: Arrays, Trees, Hypercubes*, Morgan-Kauffman, San Mateo, CA, 1992.

[182] F.T. Leighton and Satish Rao, An approximate max-flow min-cut theorem for uniform multicommodity flow problem with applications to approximation algorithms, *29nd Symposium on Foundations of Computer Science*, IEEE Computer Society Press, (1988), 422-431.

[183] T. Lengauer and R.E. Tarjan, Asymptotically tight bounds on time-space trade-offs in a pebble game, *J. Assoc. Comput. Mach.* **29** (1982), 1087-1130.

[184] H. Lenstra, personal communication.

[185] G. Lev, Size bounds and parallel algorithms for networks, Ph.D. Thesis, Department of Computer Science, University of Edinburgh.

[186] W. Li, Character sums and abelian Ramanujan graphs, *J. Number Theory* **41** (1992), 199-217.

[187] P. Li and S. T. Yau, On the parabolic kernel of the Schrödinger operator, *Acta Mathematica* **156** (1986), 153-201.

[188] P. Li and S. T. Yau, Estimates of eigenvalues of a compact Riemannian manifold, *Amer. Math. Soc. Proc. Symp. Pure Math.* **36** (1980), 205-240.

[189] J.H. Lindsey, Assignment of numbers to vertices, *Amer. Math. Monthly* **71** (1964), 508-516.

[190] R.J. Lipton and R.E. Tarjan, A separator theorem for planar graphs, *SIAM J. Appl. Math.* **36** (1979), 177-189.

[191] L. Lovász, On the Shannon capacity of a graph, *Transactions on Information Theory* -**IT-25**, IEEE Computer Society Press, (1979), 1-7.

[192] L. Lovász, Perfect graphs, in *Selected Topics in Graph Theory* **2**, (eds R. L. Wilson and L.W. Beineke), Academic Press, New York (1983), 55-87.

[193] L. Lovász and M. Simonovits, Random walks in a convex body and an improved volume algorithm, *Random Structures and Algorithms* **4** (1993), 359-412.

[194] A. Lubotzky, R. Phillips and P. Sarnak, Ramanujan graphs, *Combinatorica* **8** (1988), 261-278.

[195] G.A. Margulis, Explicit constructions of concentrators, *Problemy Peredaci Informacii* **9** (1973), 71-80 (English transl. in *Problems Inform. Transmission* **9** (1975), 325-332.

[196] G.A. Margulis, Arithmetic groups and graphs without short cycles, *6th Internat. Symp. on Information Theory, Tashkent* (1984) *Abstracts* **1**, 123-125 (in Russian).

[197] G.A. Margulis, Some new constructions of low-density parity check codes, *3rd Internat. Seminar on Information Theory, convolution codes and multi-user communication, Sochi* (1987), 275-279 (in Russian).

[198] G.A. Margulis, Explicit group theoretic constructions of combinatorial schemes and their applications for the construction of expanders and concentrators, *Problemy Peredaci Informacii* (1988) (in Russian).

[199] J.C. Maxwell, *A Treatise on Electricity and Magnetism I*, Oxford, Clarendon Press (1892), 403-410.

[200] B. Mohar, Isoperimetric number of graphs, *J. of Comb. Theory (B)* **47** (1989), 274-291.

[201] R.J. McEliece, E. R. Rodemich, and H.C. Rumsey, Jr., The Lovász and some generalizations, *J. Combinatorics, Inform. Syst. Sci.* **3** (1978), 134-152.

[202] H.L. Montgomery, *Topics in Multiplicative Number Theory*, Lecture Notes in Math. **227**, Springer-Verlag, New York (1971).

[203] J. W. Moon, *Counting Labelled Trees*, Canadian Mathematical Monographs, Canadian Mathematical Congress, Montreal, 1970.

[204] Zs. Nagy, A constructive estimation of the Ramsey numbers, *Mat. Lapok* **23** (1975), 301-302.

[205] A. Nilli, On the second eigenvalue of a graph, *Discrete Math.* **91** (1991), 207-210.

[206] B. Osgood, R. Phillips, and P. Sarnak, Extremals of determinants of Laplacians, *J. Funct. Anal.* **80** (1988), 148-211.

[207] B. Osgood, R. Phillips, and P. Sarnak, Moduli space, heights and isospectral sets of plane domains, *Ann. Math.* **129** (1989), 293-362.

[208] E.M. Palmer, *Graphical Evolution*, John Wiley and Sons, New York (1985).

[209] W.J. Paul, R.E. Tarjan and J.R. Celoni, Space bounds for a game on graphs, *Math. Soc. Theory* **10** (1977), 239-251.

[210] O. Perron, Zur Theorie der Matrizen, *Math. Ann.* **64** (1907), 248-263.

[211] M. Pinsker, On the complexity of a concentrator, 7th Internat. Teletraffic Conf., Stockholm, June 1973, 318/1-318/4.

[212] N. Pippenger, Superconcentrators, *SIAM J. Comput.* **6** (1977), 298-304.

[213] N. Pippenger, Advances in pebbling, *Internat. Collo. On Automation Languages and Programming* **9** (1982), 407-417.

[214] G. Polyá and S. Szegö, Isoperimetric Inequalities in Mathematical Physics, *Annals of Math. Studies*, no. 27, Princeton University Press, (1951).

[215] A.M. Polyakov, Quantum geometry of bosonic strings, *Phys. Letters B* **103** (1981), 207-210.

[216] P. Sarnak, *Some Applications of Modular Forms*, Cambridge University Press, Cambridge, 1990.

[217] S. Ramanujan, On certain arithmetical functions, *Trans. Cambridge Philos. Scc.* **22** (9) (1916), 159-184.

[218] F.P. Ramsey, On a problem of formal logic, *Proc. London Math. Soc.* **30** (1930), 264-286.

[219] D. Ray and I.M. Singer, Analytic torsion for complex manifolds, *Ann. Math.* **98** (1973), 154.

[220] D.K. Ray-Chaudihuri and R.M. Wilson, On t-designs, *Osaka J. Math.* **12** (1975), 745-744.

[221] A. Rényi, On the enumeration of trees, in *Combinatorial Structures and their Applications* (R. Guy, H. Hanani, N. Sauer, and J. Schonheim, eds.) Gordon and Breach, New York (1970), 355-360.

[222] V. Rödl, On the universality of graphs with uniformly distributed edges, *Discrete Math.* **59** (1986), 125-134.

[223] P. Sarnak, Determinants of Laplacians, *Comm. Math. Phys.* **110** (1987), 113-120.

[224] A. Schrijver, A comparison of the Delsarte and Lovász bounds, *IEEE Transactions on Information Theory* **IT-25**, (1979),425-429.

[225] J.J. Seidel, Graphs and their spectra, *Combinatorics and Graph Theory*, PWN-Polish Scientific Publishers, Warsaw (1989), 147-162.

[226] R. Seidel, On the all-pairs-shortest-path problem, *Proc. Sym. Theo. on Computing*, ACM (1992), 745-749.

[227] M. Simonovits and V.T. Sós, Szemerédi partitions and quasi-randomness, *Random Structures and Algorithms* **2** (1991), 1-10.

[228] A.J. Sinclair, *Algorithms for Random Generation and Counting*, Birkhauser, Boston, 1993.

[229] A.J. Sinclair and M.R. Jerrum, Approximate counting, uniform generation, and rapidly mixing markov chains, *Information and Computation* 82 (1989), 93-133.

[230] R. Singleton, On minimal graphs of maximum even girth, *J. Combin. Theory* **1** (1966), 306-332.

[231] J. Spencer, Optimal ranking of tournaments, *Networks* **1** (1971), 135-138.

[232] J. Spencer, Ramsey's theorem – A new lower bound, *J. Combinatorial Theory* **18** (1975), 108-115.

[233] S. Sternberg, *Group Theory in Physics*, Cambridge University Press (1992).

[234] D.W. Stroock, Logarithmic Sobolev inequalities for Gibbs states, Springer Lecture Notes 1563, Berlin, (1993).

[235] J. J. Sylvester, On the change of systems of independent variables, *Quarterly Journal of Mathematics* **1** (1857), 42-56. Collected Mathematical Papers, Cambridge 2 (1908), 65-85.

[236] R.M. Tanner, Explicit construction of concentrators from generalized N-gons, *SIAM J. Algebraic Discrete Methods* **5** (1984), 287-294.

[237] A. Thomason, Random graphs, strongly regular graphs and pseudo-random graphs, in Survey in Combinatorics (1987) (C. Whitehead, ed.), London Math. Soc., 173-195.

[238] A. Thomason, Pseudo-random graphs, Proc. Random Graphs, Poznán (1985) (M. Karónski, ed.), *Annals of Discrete Math.* **33** (1987), 307-331.

[239] A. Thomason, Dense expanders and pseudo-random bipartite graphs, *Discrete Math.* **75** (1989), 381-386.

[240] M. Tompa, Time space tradeoffs for computing using connectivity properties of the circuits, *J. Comput. System Sci.* **10** (1980), 118-132.

[241] H.M. Trent, Note on the enumeration and listing of all possible trees in a connected linear graph, *Proceedings of the National Academy of Sciences, U.S.A.* **40** (1954), 1004-1007.

[242] W.T. Tutte, *Graph Theory*, Addison Wesley, Reading, MA 1984.

[243] G.E. Uhlenbeck and G.W. Ford, *Lectures in Statistical Mechanics*, Providence, American Mathematical Society (1963)

[244] L. G. Valiant, The complexity of computing the permanent, *Theoret. Comput. Sci.* **8** (1979), 189-201.

[245] L.G. Valiant, Graph theoretic properties in computational complexity, *J. Comput. System Sci.* **13** (1976), 278-285.

[246] L.G. Valiant, A scheme for fast parallel communication, *SIAM J. Comput.* **11** (1982), 350-361.

[247] N.T. Varopoulos, Isoperimetric inequalities and Markov chains, *J. Funct. Anal.* **63** (1985), 215-239.

[248] K. Vijayan and U.S.R. Murty, On accessibility in graphs, *Sankhya Ser. A.* **26** (1964), 299-302.

[249] D.-L. Wang and P. Wang, Discrete isoperimetric problems, *SIAM J. Appl. Math.* **32** (1977), 860-870.

[250] A. Weil, Sur les courbes algébrique et les variétés qui s'en déduisent, *Actualités Sci. Ind.* No. 1041 (1948).

[251] H. S. Wilf, The eigenvalues of a graph and its chromatic number, *J. London Math. Soc.* **42** (1967), 330–332.

[252] R. M. Wilson, Constructions and uses of pairwise balanced designs in Combinatorics (M. Hall, Jr., and J.H. van Lint, eds.) Math Centre Tracts **55**, Amsterdam (1974), 18-41.

[253] A. Weiss, Girths of bipartite sextet graphs, *Combinatorica* **4** (1984), 241-245.

[254] S. T. Yau and Richard M. Schoen, *Differential Geometry*, Science Publications, Beijing, 1988 (Chinese version); International Press, Cambridge, Massachusetts, 1994 (English version).

[255] Louxin Zhang, Optimal bounds for matching routing on trees, *Proc. The Eighth Annual ACM-SIAM Symposium on Discrete Algorithms*, New Orleans, (1997), 445-453.

Index

$(\pi; p)$-norm, 185
L_2-norm, 15
χ-squared distance, 19
k-access graph, 102
n-cube, 20, 37
p-norm, 41

abelian group, 140
adjacency matrix, 3
adjacent, 3
all distances algorithm, 44
all shortest paths algorithm, 44
almost regular, 82
aperiodicity, 14
automorphism, 20
automorphism group, 114

bipartite, 6, 7
bipartite expander graph, 96
bipartite graph, 96
bond, 122
boundary conditions, 179
boundary expansion property, 141
boundary operator, 3
Buckyball, 122
Buckyball graph, 113

cartesian product, 37, 66
Cayley graphs, 114
chain, 3
Chebyshev polynomial, 44, 46
Cheeger constant, 24, 25, 60, 93, 115
  characterization, 32
  modified, 35, 36
  weighted graphs, 36
Cheeger inequality, 26
chromatic number, 108

clique, 81, 91, 110
clique number, 110
co-NP-complete, 94
coboundary operator, 3
comparison theorems, 68
complete graph, 6
concentrator, 93
conductance, 94
connected, 7, 24
consistent, 36
contingency table problem, 160
contraction, 13
convex subgraph, 141, 154
coset graph, 99
cycle, 6

degree, 2
deviation, 82
diameter, 8, 43, 106
diameter algorithm, 44
diameter-eigenvalue inequalities, 44
Dirichlet boundary condition, 127, 151, 194
Dirichlet eigenvalues, 127, 132–134, 136, 154
Dirichlet sum, 4
discrepancy, 75, 78, 86
distance, 8, 45
distance transitive, 113, 118, 120

edge boundary, 24, 127
edge expansion, 25
edge generating set, 114, 155
edge transitive, 116
edge-cut, 23
edge-transitive, 114
eigenfunction, 4
  harmonic, 4
eigenvalue, 4

entropy function, 105
ergodic, 14
expander coefficient, 93
expander graph, 91–93
expansion factor, 93
explicit construction, 91, 97

flow, 59
Fourier coefficients, 45

girth, 107
graph, 2
  weighted undirected, 12
group representation theory, 121

Hamiltonian path, 59
Hamming balls, 37
Harnack inequality, 139, 142
heat equation, 149
heat kernel, 149
heat kernel eigenvalue inequality, 152
homogenous graphs, 123
Hooke's law, 124
hypercube, 37, 199

independent set, 81
induced subgraph, 127
inner product, 4
intersection graph, 121
invariant, 140
irreducibility, 14
isoperimetric dimension, 167, 183
isoperimetric inequalities, 169
isoperimetric number, 35, 94
isotropy group, 114

Laplace operator, 50, 53, 157
Laplace-Beltrami operator, 4
Laplacian, 2
  determinant of , 135
lattice graphs, 155
lazy walk, 16, 18, 181
local k-frame, 199
log-Sobolev constant, 181, 187
  with Dirichlet boundary condition, 183
  with Neumann boundary condition, 183
logarithmic Harnack inequalities, 182
logarithmic Sobolev inequalities, 22, 184
loop, 2

major access network, 102
Margulis graph, 99
matching, 104, 106
matrix multiplication, 44
matrix-tree theorem, 133
measure, 4, 24
moderate growth rate, 169
Moore bound, 106

neighborhood, 74, 93, 94, 199

Neumann boundary condition, 127, 150, 194
Neumann eigenvalues, 127, 130, 154
non-blocking network, 102
normal, 91
NP, 112
NP-complete, 59

odd-even transposition sort, 65
orthonormal labeling, 111

pairing, 67
Paley graph, 81, 97
Paley sum graph, 98
partition function, 135
path, 6
perfect graph, 110
Petersen graph, 121
Plancherel formula, 20
Poincaré inequalities, 68
polynomial growth rate , 169
probabilistic method, 91
projection, 149

quasi-random, 73
quaternion group, 101

Raman spectrum, 125
Ramanujan graph, 44, 101, 107
Ramsey property, 80
random graph, 112
random walk, 14, 62
  cartesian product of graphs, 37
  ergodic, 14
  irreducible, 14
  modified, 16
  Neumann, 130
  reversible, 14
Rayleigh quotient, 4
regular, 3
relative pointwise distance, 17, 165, 181
Ricci flat, 199
Riemannian manifold, 4
rooted spanning forest, 133, 136
route covering number, 67
route set, 59
routing, 59
  complete bipartite graph, 65
  complete graph, 65
  hypercube, 65, 67
  tree, 65
routing assignment, 64
routing number, 64

separator, 23
size, 23
Sobolev constant, 41, 167, 168
Sobolev inequalities, 167
spanning tree, 133
spectrum, 4
star, 6

stationary distribution, 14
strongly convex subgraph, 140, 194
support, 51
symmetric group, 64

total variation distance, 18
trace, 10
transpose, 3, 17
transposition, 64

vertex boundary, 24, 127
vertex expansion, 24
vertex transitive, 113, 114, 117
vertex weight, 4
vertex-cut, 23
vertex-transitive, 114
vertex-transitive graph, 20
vibrational Laplacian, 123
vibrational spectrum, 123
volume, 5
Voronoi region, 158

walk, 14
weight function, 36
weighted cartesian product, 184
weighted graph, 165, 179

Selected Titles in This Series

(*Continued from the front of this publication*)

56 **Hari Bercovici, Ciprian Foiaş, and Carl Pearcy,** Dual algebras with applications to invariant subspaces and dilation theory, 1985

55 **William Arveson,** Ten lectures on operator algebras, 1984

54 **William Fulton,** Introduction to intersection theory in algebraic geometry, 1984

53 **Wilhelm Klingenberg,** Closed geodesics on Riemannian manifolds, 1983

52 **Tsit-Yuen Lam,** Orderings, valuations and quadratic forms, 1983

51 **Masamichi Takesaki,** Structure of factors and automorphism groups, 1983

50 **James Eells and Luc Lemaire,** Selected topics in harmonic maps, 1983

49 **John M. Franks,** Homology and dynamical systems, 1982

48 **W. Stephen Wilson,** Brown-Peterson homology: an introduction and sampler, 1982

47 **Jack K. Hale,** Topics in dynamic bifurcation theory, 1981

46 **Edward G. Effros,** Dimensions and C^*-algebras, 1981

45 **Ronald L. Graham,** Rudiments of Ramsey theory, 1981

44 **Phillip A. Griffiths,** An introduction to the theory of special divisors on algebraic curves, 1980

43 **William Jaco,** Lectures on three-manifold topology, 1980

42 **Jean Dieudonné,** Special functions and linear representations of Lie groups, 1980

41 **D. J. Newman,** Approximation with rational functions, 1979

40 **Jean Mawhin,** Topological degree methods in nonlinear boundary value problems, 1979

39 **George Lusztig,** Representations of finite Chevalley groups, 1978

38 **Charles Conley,** Isolated invariant sets and the Morse index, 1978

37 **Masayoshi Nagata,** Polynomial rings and affine spaces, 1978

36 **Carl M. Pearcy,** Some recent developments in operator theory, 1978

35 **R. Bowen,** On Axiom A diffeomorphisms, 1978

34 **L. Auslander,** Lecture notes on nil-theta functions, 1977

33 **G. Glauberman,** Factorizations in local subgroups of finite groups, 1977

32 **W. M. Schmidt,** Small fractional parts of polynomials, 1977

31 **R. R. Coifman and G. Weiss,** Transference methods in analysis, 1977

30 **A. Pełczyński,** Banach spaces of analytic functions and absolutely summing operators, 1977

29 **A. Weinstein,** Lectures on symplectic manifolds, 1977

28 **T. A. Chapman,** Lectures on Hilbert cube manifolds, 1976

27 **H. Blaine Lawson, Jr.,** The quantitative theory of foliations, 1977

26 **I. Reiner,** Class groups and Picard groups of group rings and orders, 1976

25 **K. W. Gruenberg,** Relation modules of finite groups, 1976

24 **M. Hochster,** Topics in the homological theory of modules over commutative rings, 1975

23 **M. E. Rudin,** Lectures on set theoretic topology, 1975

22 **O. T. O'Meara,** Lectures on linear groups, 1974

21 **W. Stoll,** Holomorphic functions of finite order in several complex variables, 1974

20 **H. Bass,** Introduction to some methods of algebraic K-theory, 1974

19 **B. Sz.-Nagy,** Unitary dilations of Hilbert space operators and related topics, 1974

18 **A. Friedman,** Differential games, 1974

17 **L. Nirenberg,** Lectures on linear partial differential equations, 1973

16 **J. L. Taylor,** Measure algebras, 1973

15 **R. G. Douglas,** Banach algebra techniques in the theory of Toeplitz operators, 1973

(See the AMS catalog for earlier titles)